The Department of Agriculture

KNOW YOUR GOVERNMENT

The Department of Agriculture

R. Douglas Hurt

CHELSEA HOUSE PUBLISHERS

Chelsea House Publishers
Editor-in-Chief: Nancy Toff
Executive Editor: Remmel T. Nunn
Managing Editor: Karyn Gullen Browne
Copy Chief: Juliann Barbato
Picture Editor: Adrian G. Allen
Art Director: Maria Epes
Manufacturing Manager: Gerald Levine

Know Your Government
Senior Editor: Kathy Kuhtz

Staff for THE DEPARTMENT OF AGRICULTURE
Deputy Copy Chief: Ellen Scordato
Editorial Assistant: Theodore Keyes
Picture Research: Dixon and Turner Research Associates
Assistant Art Director: Laurie Jewell
Designer: Noreen M. Lamb
Production Coordinator: Joseph Romano

First Printing

1   3   5   7   9   8   6   4   2

Library of Congress Cataloging in Publication Data
Hurt, R. Douglas.
  The Department of Agriculture.
  (Know your government)
  Bibliography: p.
  Includes index.
  Summary: Surveys the history of the Department of Agriculture, describing its
structure, current function, and influence on American society.
  1. United States. Dept. of Agriculture. [1. United States. Dept. of Agriculture] I.
Title. II. Series: Know your government (New York, N.Y.)
S21.C9H87 1989   353.81'09   88-18139
ISBN 0-87754-833-1
      0-7910-0854-1 (pbk.)

CONTENTS

KNOW YOUR GOVERNMENT

CHELSEA HOUSE PUBLISHERS

Government: Crises of Confidence

Arthur M. Schlesinger, jr.

From the start, Americans have regarded their government with a mixture of reliance and mistrust. The men who founded the republic did not doubt the indispensability of government. "If men were angels," observed the 51st Federalist Paper, "no government would be necessary." But men are not angels. Because human beings are subject to wicked as well as to noble impulses, government was deemed essential to assure freedom and order.

At the same time, the American revolutionaries knew that government could also become a source of injury and oppression. The men who gathered in Philadelphia in 1787 to write the Constitution therefore had two purposes in mind. They wanted to establish a strong central authority and to limit that central authority's capacity to abuse its power.

To prevent the abuse of power, the Founding Fathers wrote two basic principles into the new Constitution. The principle of federalism divided power between the state governments and the central authority. The principle of the separation of powers subdivided the central authority itself into three branches—the executive, the legislative, and the judiciary—so that "each may be a check on the other." The *Know Your Government* series focuses on the major executive departments and agencies in these branches of the federal government.

7

The Constitution did not plan the executive branch in any detail. After vesting the executive power in the president, it assumed the existence of "executive departments" without specifying what these departments should be. Congress began defining their functions in 1789 by creating the Departments of State, Treasury, and War. The secretaries in charge of these departments made up President Washington's first cabinet. Congress also provided for a legal officer, and President Washington soon invited the attorney general, as he was called, to attend cabinet meetings. As need required, Congress created more executive departments.

Setting up the cabinet was only the first step in organizing the American state. With almost no guidance from the Constitution, President Washington, seconded by Alexander Hamilton, his brilliant secretary of the treasury, equipped the infant republic with a working administrative structure. The Federalists believed in both executive energy and executive accountability and set high standards for public appointments. The Jeffersonian opposition had less faith in strong government and preferred local government to the central authority. But when Jefferson himself became president in 1801, although he set out to change the direction of policy, he found no reason to alter the framework the Federalists had erected.

By 1801 there were about 3,000 federal civilian employees in a nation of a little more than 5 million people. Growth in territory and population steadily enlarged national responsibilities. Thirty years later, when Jackson was president, there were more than 11,000 government workers in a nation of 13 million. The federal establishment was increasing at a faster rate than the population.

Jackson's presidency brought significant changes in the federal service. He believed that the executive branch contained too many officials who saw their jobs as "species of property" and as "a means of promoting individual interest." Against the idea of a permanent service based on life tenure, Jackson argued for the periodic redistribution of federal offices, contending that this was the democratic way and that official duties could be made "so plain and simple that men of intelligence may readily qualify themselves for their performance." He called this policy rotation-in-office. His opponents called it the spoils system.

In fact, partisan legend exaggerated the extent of Jackson's removals. More than 80 percent of federal officeholders retained their jobs. Jackson discharged no larger a proportion of government workers than Jefferson had done a generation earlier. But the rise in these years of mass political parties gave federal patronage new importance as a means of building the party and of rewarding activists. Jackson's successors were less restrained in the distribution of spoils. As the federal establishment grew—to nearly 40,000 by 1861—the politicization of the public service excited increasing concern.

After the Civil War the spoils system became a major political issue. High-minded men condemned it as the root of all political evil. The spoilsmen, said the British commentator James Bryce, "have distorted and depraved the mechanism of politics." Patronage, by giving jobs to unqualified, incompetent, and dishonest persons, lowered the standards of public service and nourished corrupt political machines. Office-seekers pursued presidents and cabinet secretaries without mercy. "Patronage," said Ulysses S. Grant after his presidency, "is the bane of the presidential office." "Every time I appoint someone to office," said another political leader, "I make a hundred enemies and one ingrate." George William Curtis, the president of the National Civil Service Reform League, summed up the indictment. He said,

> The theory which perverts public trusts into party spoils, making public
> employment dependent upon personal favor and not on proved merit,
> necessarily ruins the self-respect of public employees, destroys the
> function of party in a republic, prostitutes elections into a desperate
> strife for personal profit, and degrades the national character by lower-
> ing the moral tone and standard of the country.

The object of civil service reform was to promote efficiency and honesty in the public service and to bring about the ethical regeneration of public life. Over bitter opposition from politicians, the reformers in 1883 passed the Pendleton Act, establishing a bipartisan Civil Service Commission, competitive examinations, and appointment on merit. The Pendleton Act also gave the president authority to extend by executive order the number of "classified" jobs—that is, jobs subject to the merit system. The act applied initially only to about 14,000 of the more than 100,000 federal positions. But by the end of the 19th century 40 percent of federal jobs had moved into the classified category.

Civil service reform was in part a response to the growing complexity of American life. As society grew more organized and problems more technical, official duties were no longer so plain and simple that any person of intelligence could perform them. In public service, as in other areas, the all-round man was yielding ground to the expert, the amateur to the professional. The excesses of the spoils system thus provoked the counter-ideal of scientific public administration, separate from politics and, as far as possible, insulated against it.

The cult of the expert, however, had its own excesses. The idea that administration could be divorced from policy was an illusion. And in the realm of policy, the expert, however much segregated from partisan politics, can never attain perfect objectivity. He remains the prisoner of his own set of values. It is these values rather than technical expertise that determine fundamental judgments of public policy. To turn over such judgments to experts, moreover, would be to abandon democracy itself; for in a democracy final decisions must be made by the people and their elected representatives. "The business of the expert," the British political scientist Harold Laski rightly said, "is to be on tap and not on top."

Politics, however, were deeply ingrained in American folkways. This meant intermittent tension between the presidential government, elected every four years by the people, and the permanent government, which saw presidents come and go while it went on forever. Sometimes the permanent government knew better than its political masters; sometimes it opposed or sabotaged valuable new initiatives. In the end a strong president with effective cabinet secretaries could make the permanent government responsive to presidential purpose, but it was often an exasperating struggle.

The struggle within the executive branch was less important, however, than the growing impatience with bureaucracy in society as a whole. The 20th century saw a considerable expansion of the federal establishment. The Great Depression and the New Deal led the national government to take on a variety of new responsibilities. The New Deal extended the federal regulatory apparatus. By 1940, in a nation of 130 million people, the number of federal workers for the first time passed the 1 million mark. The Second World War brought federal civilian employment to 3.8 million in 1945. With peace, the federal establishment declined to around 2 million by 1950. Then growth resumed, reaching 2.8 million by the 1980s.

The New Deal years saw rising criticism of "big government" and "bureaucracy." Businessmen resented federal regulation. Conservatives worried about the impact of paternalistic government on individual self-reliance, on community responsibility, and on economic and personal freedom. The nation in effect renewed the old debate between Hamilton and Jefferson in the early republic, although with an ironic exchange of positions. For the Hamiltonian constituency, the "rich and well-born," once the advocate of affirmative government, now condemned government intervention, while the Jeffersonian constituency, the plain people, once the advocate of a weak central government and of states' rights, now favored government intervention.

In the 1980s, with the presidency of Ronald Reagan, the debate has burst out with unusual intensity. According to conservatives, government intervention abridges liberty, stifles enterprise, and is inefficient, wasteful, and arbitrary. It disturbs the harmony of the self-adjusting market and creates worse troubles than it solves. Get government off our backs, according to the popular cliché, and our problems will solve themselves. When government is necessary, let it be at the local level, close to the people. Above all, stop the inexorable growth of the federal government.

In fact, for all the talk about the "swollen" and "bloated" bureaucracy, the federal establishment has not been growing as inexorably as many Americans seem to believe. In 1949, it consisted of 2.1 million people. Thirty years later, while the country had grown by 70 million, the federal force had grown only by 750,000. Federal workers were a smaller percentage of the population in 1985 than they were in 1955—or in 1940. The federal establishment, in short, has not kept pace with population growth. Moreover, national defense and the postal service account for 60 percent of federal employment.

Why then the widespread idea about the remorseless growth of government? It is partly because in the 1960s the national government assumed new and intrusive functions: affirmative action in civil rights, environmental protection, safety and health in the workplace, community organization, legal aid to the poor. Although this enlargement of the federal regulatory role was accompanied by marked growth in the size of government on all levels, the expansion has taken place primarily in state and local government. Whereas the federal force increased by only 27 percent in the 30 years after 1950, the state and local government force increased by an astonishing 212 percent.

Despite the statistics, the conviction flourishes in some minds that the national government is a steadily growing behemoth swallowing up the liberties of the people. The foes of Washington prefer local government, feeling it is closer to the people and therefore allegedly more responsive to popular needs. Obviously there is a great deal to be said for settling local questions locally. But local government is characteristically the government of the locally powerful. Historically, the way the locally powerless have won their human and constitutional rights has often been through appeal to the national government. The national government has vindicated racial justice against local bigotry, defended the Bill of Rights against local vigilantism, and protected natural resources against local greed. It has civilized industry and secured the rights of labor organizations. Had the states' rights creed prevailed, there would perhaps still be slavery in the United States.

The national authority, far from diminishing the individual, has given most Americans more personal dignity and liberty than ever before. The individual freedoms destroyed by the increase in national authority have been in the main the freedom to deny black Americans their rights as citizens; the freedom to put small children to work in mills and immigrants in sweatshops; the freedom to pay starvation wages, require barbarous working hours, and permit squalid working conditions; the freedom to deceive in the sale of goods and securities; the freedom to pollute the environment—all freedoms that, one supposes, a civilized nation can readily do without.

"Statements are made," said President John F. Kennedy in 1963, "labelling the Federal Government an outsider, an intruder, an adversary. . . . The United States Government is not a stranger or not an enemy. It is the people of fifty states joining in a national effort. . . . Only a great national effort by a great people working together can explore the mysteries of space, harvest the products at the bottom of the ocean, and mobilize the human, natural, and material resources of our lands."

So an old debate continues. However, Americans are of two minds. When pollsters ask large, spacious questions—Do you think government has become too involved in your lives? Do you think government should stop regulating business?—a sizable majority opposes big government. But when asked specific questions about the practical work of government—Do you favor social security? unemployment compensation? Medicare? health and safety standards in factories? environmental protection? government guarantee of jobs for everyone seeking employment? price and wage controls when inflation threatens?—a sizable majority approves of intervention.

In general, Americans do not want less government. What they want is more efficient government. They want government to do a better job. For a time in the 1970s, with Vietnam and Watergate, Americans lost confidence in the national government. In 1964, more than three-quarters of those polled had thought the national government could be trusted to do right most of the time. By 1980 only one-quarter was prepared to offer such trust. But by 1984 trust in the federal government to manage national affairs had climbed back to 45 percent.

Bureaucracy is a term of abuse. But it is impossible to run any large organization, whether public or private, without a bureaucracy's division of labor and hierarchy of authority. And we live in a world of large organizations. Without bureaucracy modern society would collapse. The problem is not to abolish bureaucracy, but to make it flexible, efficient, and capable of innovation.

Two hundred years after the drafting of the Constitution, Americans still regard government with a mixture of reliance and mistrust—a good combination. Mistrust is the best way to keep government reliable. Informed criticism is the means of correcting governmental inefficiency, incompetence, and arbitrariness; that is, of best enabling government to play its essential role. For without government, we cannot attain the goals of the Founding Fathers. Without an understanding of government, we cannot have the informed criticism that makes government do the job right. It is the duty of every American citizen to know our government—which is what this series is all about.

The breeder hens of Perdue Inc., in Salisbury, Maryland. The USDA's veterinary services help to determine the existence and extent of diseases and pests affecting livestock and poultry.

ONE

Improving America's Farms

On May 15, 1862, President Abraham Lincoln signed the act establishing the U.S. Department of Agriculture (USDA). The law authorized the department to increase the nation's agricultural knowledge through research and to help America's farmers use the discoveries and improved agricultural practices to become more efficient, productive, and profitable.

The new department's scientists immediately began to work on many research projects, such as improving plant varieties so that they could withstand drought and disease, determining the causes of livestock diseases and finding the cures, and learning the nutritional requirements of plants and animals. Indeed, until the late 19th century, USDA scientists, technicians, and officials were dedicated primarily to making two blades of grass grow where only one grew before—that is, to making the nation's farms more productive. The department's administrators based this policy on the assumption that increased productivity would enable farmers to sell larger quantities of agricultural commodities and earn more money. As farmers increased their income and improved their standard of living, the public would also enjoy an abundant supply of food and fiber, such as cotton and wool.

During the 1880s, the department began to be responsible for the area of regulation. USDA officials took part in congressionally approved progams to

15

ensure the health and safety of livestock shipped in interstate commerce and livestock imported into the United States. They also worked to prohibit the export of diseased livestock and livestock products. In addition, to ensure that consumers purchased only wholesome food products, the USDA investigated the adulteration of foods (making food impure by the addition of a foreign or inferior substance, or the substitution for or removal of an important ingredient) by manufacturers.

Although the USDA's earliest responsibilities—research, education, and regulation—remained important, during the first half of the 20th century, the department started to organize programs to aid farmers in the marketplace and to increase their purchasing power. After passage of the Agricultural Adjustment Act of 1933, the department paid farmers to limit the production of wheat, cotton, corn, rice, tobacco, and milk and the raising of swine to keep prices at profitable levels. By the late 20th century, price and income support programs were among the most important services that the department provided to farmers.

The department continues to serve as a major research and educational institution. It works closely with land-grant colleges (state agricultural and mechanical colleges, built on land granted by the federal government, that teach agriculture, engineering, and home economics) and state experiment stations to increase and share scientific discoveries in agriculture that will improve farming and rural life. USDA employees work with the Cooperative Extension System (a nationwide educational network that joins research, science, and technology to people's needs) to supply agents and home economists to nearly every county in the United States. These individuals share the results of new agricultural developments with farmers and other members of the public to help make their lives more productive and enjoyable.

The Department of Agriculture also continues to be the leading regulatory agency in the federal government. It administers more legislation and programs—designed to protect the interests of farmers, food manufacturers, and consumers—than any other federal agency. Consequently, it helps safeguard the public's health and prevents fraud and unfair business practices in the marketing of agricultural commodities and food products. The department's marketing and planting regulations also help to ensure dependable food supplies at reasonable prices for the consumer.

Internationally, the USDA serves as the American farmers' representative for the development of new markets around the world. The department's employees, called attachés, are stationed at the nation's foreign embassies and consulates (offices that conduct day-to-day political and economic relations

Strip-cropping on a farm in northwestern Illinois. Growing a culti-vated crop, such as corn, in strips and alternating it with strips of another crop, such as alfalfa, protects the land from excess runoff and soil erosion. USDA extension agents work closely with farmers to conserve nutrients in the soil.

with other nations, promote commercial interests, and assist Americans abroad). They gather information that will help farmers and shippers sell agricultural commodities in international commerce. The department is also responsible for sending surplus food commodities to needy nations—Ethiopia, for example—under the authorization of Public Law 480, also known as the Food for Peace program. Congress established this program in 1954 to reduce the nation's surpluses of food, such as wheat, and to increase consumption of American goods abroad. In addition, the department plays a major role in helping developing nations use their agricultural resources to their fullest potential. The department sends scientists abroad to teach improved agricul-tural practices to nations that seek American help, and it invites foreign scientists to the United States to study American agriculture and farming methods.

The USDA does not help only farmers, however; it also delivers services and programs that benefit rural residents who do not live on farms and those who live in cities far removed from the fields and pastures of the countryside. For example, the USDA provides loans for the construction of low-cost rural housing and for the support of cooperatives that furnish electricity to rural areas. The department also encourages good nutrition for people with low incomes by means of a food stamp program, enabling them to purchase groceries to prepare balanced meals. The USDA distributes surplus food, such as cheese, butter, rice, honey, and non-fat dry milk, to people in need. Moreover, it supports breakfast and lunch programs for the nation's school-children. By collecting and analyzing agricultural statistics both at home and abroad, the Department of Agriculture provides the information necessary for policymakers to respond to agricultural problems—for instance, those involving supply and demand—and to help improve the quality of rural and urban life. The USDA also strives to conserve the nation's soil and forests so that future generations will have adequate food, timber, water, and recreational areas.

Workers at a fish hatchery at the College of Southern Idaho carry out research on genetic breeding to improve fish quality. The USDA provides funding and expert advice for college-run hatcheries such as this.

Indeed, the USDA's Soil Conservation Service and the Forest Service are two of its most prominent agencies.

Although the USDA's headquarters is located in Washington, D.C., approximately 90 percent of the department's more than 112,600 employees do not work in the nation's capital, but are scattered across the country. USDA employees can be found in every state and nearly every county in the nation. Some are extension agents (employees who educate the public, providing the results of the latest agricultural research), and a number are home economists (specialists in such fields as nutrition, clothing, and child care). Others administer crop production and soil conservation programs. Some USDA employees safeguard our national forests, and others conduct scientific research to determine new uses for food products or to improve agricultural productivity. Through the work of these men and women, the Department of Agriculture touches the lives of every resident of the United States every day.

An 1802 award commemorated an American consul's endeavors to import an improved breed of sheep from Spain to New England. The Massachusetts Society for Promoting Agriculture was one of the earliest agricultural societies in America.

TWO

Experimentation and Expansion 1613–1862

Before the development of agricultural colleges and agricultural experiment stations during the late 19th century, few farmers had the information necessary to improve their lands, crops, and livestock. How should a farmer in Georgia restore worn-out soil? How could grape farmers in New York prevent disease in their vineyards? How could wheat farmers in Pennsylvania halt the loss of their crops to insects? In the absence of authoritative, scientific information that was readily available, most farmers relied upon common sense or superstition or tradition to solve their problems. But often their solutions were inadequate to meet the difficulties or the needs at hand. There was no federal agency that could provide them with the information they needed to improve productivity and to increase profits.

But farmers have always experimented with various techniques to improve their crops and to make their work more efficient and profitable. As early as 1613, John Rolfe experimented with Orinoco tobacco (a type of tobacco from South America), near Jamestown, Virginia—the first permanent English settlement in America—and in 1669, some farmers in South Carolina experimented with various tropical crops, such as cotton, indigo, ginger, sugarcane, and olives. In 1733, leaders in the colony of Georgia established an experimental garden and hired a botanist to collect plants from the West Indies and Central and South America.

These efforts to gain knowledge about agriculture, however, did little to enhance farming in the American colonies because crops such as oranges, figs, and olives were unsuited to the climate of the areas where the experiments were conducted. Even so, rapid agricultural change in Great Britain and the dependence of most Americans on farming for a livelihood encouraged them to learn more about agriculture to improve their standard of living. During the 18th century, farmers, including such notables as George Washington and Thomas Jefferson, corresponded with British agricultural leaders, including Arthur Young and Sir John Sinclair—both of whom had written extensively on agriculture and economics. Americans wanted to learn about British advances in farming in order to apply that knowledge to their own plantations.

Another significant development that improved the accuracy and increased the amount of agricultural information available to American farmers during the 18th century was the formation of agricultural societies by wealthy farmers and planters. Members of these agricultural societies shared information and conducted experiments—for example, those designed to improve soil fertility and to exterminate insects. These wealthy farmers had the leisure to undertake agricultural experiments. Moreover, they had the ability and willingness to share their discoveries with others and lectured, corresponded, and published articles in agricultural pamphlets and magazines, such as *Agricultural Enquiries on Plaster of Paris* and the *Pennsylvania Mercury and Universal Advertiser*. They also shared the results of their experiments with members at society meetings. In 1785, the Philadelphia Society for Promoting Agriculture and the South Carolina Society for Promoting Agriculture and Other Rural Concerns were the first two such groups organized to support and share the discovery of agricultural research.

These organizations, together with the experimental efforts of individuals such as Eli Whitney and Thomas Jefferson, as well as acts of Congress, such as the Land Ordinance of 1785, helped to encourage agricultural improvement and expansion in the United States. In 1793, for example, Eli Whitney developed a machine, called a cotton gin, that efficiently removed the seeds and hulls from the cotton fibers. Because the cotton gin could do as much work mechanically in one day as 50 people could complete working by hand, the cotton gin enabled farmers to plant and harvest more cotton and get it to market, thereby making cotton farming profitable in the United States. Several years later, in 1798, Thomas Jefferson demonstrated that the moldboard of a plow could be designed according to mathematical principles to enable standardization and easy reproduction as well as efficient plowing. Moreover, the Land Ordinance of 1785 helped open the Old Northwest (today the states of Ohio, Indiana, Illinois, Wisconsin, and Michigan) to settlement and agriculture by providing for the orderly survey and sale of the land in that area. Developments such as these helped to make the dissemination of agricultural knowledge increasingly important.

After Great Britain established a Board of Agriculture in 1793 to survey rural conditions and to advise farmers on ways to improve their operations, the members of the various agricultural societies in the United States recognized that a similar governmental agency could provide valuable information to them. Agricultural leaders throughout the new nation urged Congress to create a comparable board or bureau.

George Washington was particularly interested in developing a governmental agency dedicated to improving American agriculture. On December 7, 1796, in his last annual message to Congress before he retired from the presidency, Washington asked Congress to create a national board of agriculture. The

George Washington (second from right) is depicted directing workers on his farm in this 1853 lithograph. Washington, like many 18th century American farmers, wrote to British agricultural authorities asking for advice about new farming methods and improved plant varieties.

23

Cyrus H. McCormick's grain reaper, patented in 1834, mechanized the cutting of grain. Widespread use of the machine did not occur until the Civil War, when laborers left the farms to enlist. The reaper enabled a woman or boy to perform the work of several men.

president wanted the board to collect and publish agricultural information and to encourage agricultural progress. The members of the House of Representatives expressed interest in Washington's proposal and appointed a committee to study it. In January 1797, the committee recommended a bill that provided for the creation of the American Society of Agriculture. A secretary, paid by the federal government, was to head the society. Although the House of Representatives did not vote on the bill for reasons that are not entirely clear, the bill encouraged the organization of more agricultural societies to improve farming. The societies persisted in urging Congress to establish a federal agricultural board or agency that would promote the interests of farmers and enhance their agricultural activities.

In 1817, the House of Representatives prepared another bill to establish such an agricultural board. Once again, however, the bill did not come to a vote, primarily because President James Madison and his supporters in Congress opposed it. Madison believed that a national board of agriculture would excessively expand the powers and the expense of the federal government.

While farmers waited for Congress to create a board of agriculture, they continued to organize for the promotion of better farming methods in their local areas. In 1809, for example, a group interested in farming organized the Columbia Agricultural Society located in Georgetown within the District of

Columbia. In May 1810, this organization held an agricultural fair to help farmers learn the best techniques to improve production and to increase profits. Soon thereafter, other agricultural societies began holding fairs for the promotion of agricultural improvement. In 1819, the *American Farmer* was published in Baltimore, Maryland, by John Stuart Skinner and became the leading periodical devoted to the increase and dissemination of agricultural knowledge.

During this period, however, the executive branch of the federal government encouraged farmers to import superior agricultural plants and livestock to help them become more enterprising. For instance, Henry Clay, the influential politician and owner of a plantation in Kentucky, imported the first Hereford cattle from Herefordshire, England, into the United States in 1817. In 1818, Elkanah Watson, a wealthy Massachusetts farmer, sent letters to American consuls (federal representatives who help Americans abroad and promote commercial interests there) asking them to send him seeds for distribution to farmers. This action may have prompted the federal government to pursue a similar venture. On March 26, 1819, William L. Crawford, secretary of the Treasury under President James Monroe, asked the American consuls to send the seeds from traditional plants and improved varieties in their respective countries to the customs collectors at American ports. The customs officials would then distribute the seeds to farmers around the country. Crawford and others believed this activity would enable farmers to raise new crops that would be both productive and profitable. Spurred by Crawford's efforts, the House of Representatives created a Committee on Agriculture in 1820 and the Senate formed a similar committee five years later.

Soon thereafter, on September 6, 1827, President John Quincy Adams directed Secretary of the Treasury Richard Rush to ask the American consuls abroad and the captains of naval ships to collect samples of the seeds from superior plants for distribution. The Treasury Department's plan was to create a botanical garden to experiment with the new plants and to distribute the seeds to farmers who agreed to plant them. President Adams wanted to learn how these plants would grow in the environmental conditions of the United States.

The federal government did little more to aid agriculture until 1836, when Henry Leavitt Ellsworth became commissioner of the newly created Patent Office, which registered inventions and gave inventors exclusive rights over their discoveries for a limited time. Ellsworth had a great interest in agriculture. Although his office did not have specific responsibility for promoting agriculture, on his own authority he quickly used his office to aid farmers. Under Ellsworth the Patent Office approved the designs for a host of new

Henry Leavitt Ellsworth, commissioner of the Patent Office, vigorously promoted the collection of seeds and plants by American consuls in foreign countries. He also circulated reports containing crop statistics and essays on improved farming techniques to congressmen who, in turn, gave the information to their constituents.

implements, such as John Deere's steel plow and the Pitts brothers' threshing machine in 1837, and Moses and Samuel Pennock's hoe drill for sowing grain in 1841. Ellsworth also continued to collect seeds from foreign countries through the consulates and the navy, distributing them to farmers through agricultural societies and congressmen. He also requested an appropriation from Congress to help finance these activities, which he had been paying for out of his own pocket.

Congress responded to Ellsworth's request on March 3, 1839, when it appropriated $1,000 to support the collection and distribution of seeds and the compilation and dissemination of agricultural statistics and other information relating to farming. This was the first time that the federal government provided assistance for agriculture. Ellsworth used the funds to establish an agricultural division in the Patent Office. (Congress did not make another appropriation until 1842 and annual appropriations did not begin until 1847.)

While Ellsworth struggled to gain annual appropriations from Congress, he also championed the creation of a federal bureau of agriculture and issued annual reports on the agricultural activities of the Patent Office. These reports were popular with congressmen, who gave them to their constituents to help assure those voters that Congress was interested in the farmers' welfare. Each annual report contained agricultural statistics on the production of crops—wheat and potatoes, for example—essays on improved methods of crop and livestock raising, and letters from correspondents throughout the United States who informed readers about agricultural problems nationwide, such as marketing concerns and the surplus production of cotton or other crops.

However, not everyone in Congress or among the general public believed that the agricultural work of the Patent Office was worthwhile. In 1846, Senator Ambrose Hundley Sevier of Arkansas charged that the reports were "comparatively worthless" and no more than an "accumulation of newspaper paragraphs." Senator John C. Calhoun of South Carolina denounced the agricultural work of the Patent Office as "one of the most enormous abuses under this Government." Senator Willie Person Mangum of North Carolina charged that "practical farmers turned up their noses with utter scorn and contempt" at the agricultural reports that the Patent Office compiled and distributed.

These criticisms were not strong enough, however, to stop the agricultural work of the Patent Office. By 1849, the office's annual report was so large that it was published in two parts. One volume was completely devoted to agriculture, and congressmen willingly distributed it free of charge to farmers in their districts. (The other volume contained information about patents.) As a result, the Patent Office continued to serve as the focal point for the agricultural activities of the federal government. (The Patent Office, which had been an independent bureau, was transferred to the Department of the Interior when that department was created in 1849.)

While the Patent Office struggled to conduct and expand its agricultural work, various farm groups on the state and local levels asked Congress to create a department of agriculture to support the improvement of farming nationwide. The United States Agricultural Society, the Maryland Agricultural Society, and the Massachusetts Board of Agriculture repeatedly petitioned Congress to create such an agency.

For example, the United States Agricultural Society, which was organized in 1852, adopted a resolution on February 2, 1853, that urged Congress to create a department of agriculture with cabinet status—that is, to establish an executive department led by a secretary who was an instrumental part of the president's policy-making circle. Congress, however, remained divided over this issue and did not act upon the proposal. During the next 10 years, the United States Agricultural Society continued to implore Congress to establish a department.

While the United States Agricultural Society pressed for the creation of a department of agriculture, the Patent Office increasingly had less time and fewer resources to continue its agricultural work. The burden finally became so great that on January 11, 1860, William D. Bishop, commissioner of Patents, asked Congress to relieve his office of its agricultural responsibilities. Although some government officials believed the agricultural work of the Patent Office should remain within the Department of the Interior, in 1861 Thomas G. Clemson, superintendent of the Agricultural Division of the Patent Office, stated that a department of agriculture should be free from the influences of a parent agency. However, Southern congressmen (and some Northern congressmen) who advocated states' rights—the retention of all powers for the states that the Constitution did not reserve for the national government—opposed that idea. They continued to argue that a department would increase the centralization of the federal government at the expense of state authority and individual liberty and that it would create an expensive bureaucratic nightmare.

When the Southern states seceded from the Union in early 1861, the way opened for the creation of a department of agriculture in the federal government. Abraham Lincoln, the newly elected president, and the Republican party supported the creation of a department. On December 3, 1861, President Lincoln recommended in his first annual message to Congress that such a department be established. As a result of that recommendation, Owen Lovejoy of Illinois, chairman of the House Committee on Agriculture, introduced a bill on January 7, 1862, for the establishment of the USDA.

Congress moved slowly on Lincoln's proposal and Lovejoy's bill. By the time both the House of Representatives and the Senate approved the creation of a new department, the original bill had been changed many times. In general, the final bill authorized the department to "acquire and to diffuse among the people of the United States useful information on subjects connected with agriculture in the most general and comprehensive sense of the word." Specifically, the bill empowered the department to acquire, test, and distribute new and valuable seeds and plants, to conduct "practical and scientific experiments," to collect agricultural statistics and other information, and to publish annual and other useful reports that could help improve agriculture.

Few congressmen opposed the creation of a department of agriculture in principle, because their constituents favored such an agency and because they believed it would provide a useful public service. They did, however, disagree about the organizational structure for the department and about the scope of its authority. Some, such as Senator J. F. Simmons of Rhode Island, wanted an

Grain cradlers harvest wheat in the late 1800s. Commissioner Ellsworth encouraged American inventors to improve the tools and machines that would enable farmers to become more efficient.

Congress of the United States

At the second Session

BEGUN AND HELD AT THE CITY OF WASHINGTON

in the District of Columbia

on Monday the second day of December one thousand eight hundred and sixty-one.

AN ACT to establish a Department of Agriculture.

Be It Enacted by the Senate and House of Representatives of the United States of America in Congress assembled.

That there is hereby established at the seat of government of the United States a Department of Agriculture, the general designs and duties of which shall be to acquire and to diffuse among the people of the United States useful information on subjects connected with agriculture in the most general and comprehensive sense of that word, and to procure, propagate, and distribute among the people new and valuable seeds and plants.

Sec. 2. And be it further enacted, That there shall be appointed by the President, by and with the advice and consent of the Senate, a "Commissioner of Agriculture", who shall be the chief executive officer of the Department of Agriculture, who shall hold his office by a tenure similar to that of other civil officers appointed by the President; and who shall receive for his compensation a salary of three thousand dollars per annum.

Sec. 3. And be it further enacted, That it shall be the duty of the Commissioner of Agriculture to acquire and preserve in his department all information concerning agriculture which he can obtain by means of books and correspondence and by practical and scientific experiments, (accurate records of which experiments shall be kept in his office,) by the collection of statistics and by any other appropriate means within his power; to collect, as he may be able new and valuable seeds and plants; to test, by cultivation, the value of such of them as may require such test; to propagate such as may be worthy of propagation, and to distribute them among agriculturists. He shall annually make a general report in

The first page of the act of Congress that established the Department of Agriculture. President Abraham Lincoln signed the document on May 15, 1862.

independent department led by a secretary with cabinet status. Other congressmen, such as Senator La Fayette Foster of Connecticut, merely wanted to create a bureau, fearing a department of cabinet status would promote too much centralization of authority at the expense of state governments. Some congressmen wanted to give the Department of Agriculture broad powers and others wanted its authority strictly limited.

Ultimately, the bill provided for an independent department without cabinet status headed by a commissioner who was appointed by the president. This feature of the bill was a compromise, drafted under the direction of Representative Lovejoy in the House Committee on Agriculture, between those who sought an independent agency with cabinet status and those who wanted a subordinate bureau of agriculture, with less prestige and responsibility, within the Department of the Interior. Agriculture now had representation in the federal government, but it did not have the honor and authority of cabinet status. Yet Congress hoped that by establishing an independent department with a commissioner, the agency would escape domination by politicians who held office for their own gain rather than for the farmers'.

Finally, both houses of Congress approved the bill to create a department and President Lincoln signed it into law on May 15, 1862. At a time when the very existence of the nation was threatened by the Civil War, Congress looked to the future and authorized the creation of the new department, designed to assist all farmers—those in the North and in the South.

*A meeting of the National Grange of the Patrons of Husbandry,
called Grangers, near Winchester, Illinois, in 1873. Originally established for social and educational purposes, granges became active in
politics as they sought to correct economic abuses such as exorbitant
charges for use of warehouses and railroads.*

THREE

The Department Blossoms 1862–1941

Congress created the Department of Agriculture in perilous times. With Union and Confederate armies battling near Richmond, Virginia, few people paid much attention to the news on July 1, 1862, that Isaac Newton had taken the oath of office as the first commissioner of Agriculture. Even though the nation was locked in the Civil War, Newton strove to develop a Department of Agriculture based on congressional guidelines.

Commissioner Newton, a Pennsylvania dairyman, had served as chief of the agricultural section in the Patent Office and clearly understood the needs of both the farmers and the new department. Newton quickly established seven objectives for the Department of Agriculture and immediately began work to achieve them. His objectives were (1) to collect and publish statistical information; (2) to collect and disseminate valuable plants and animals to farmers; (3) to test agricultural implements; (4) to conduct chemical analyses of soils, grains, fruits, plants, vegetables, and manures; (5) to establish a professorship for the study of botany and entomology within the department; (6) to install an agricultural library and museum; and (7) to serve as the farmers' most important adviser on all agricultural matters.

Under Newton, the Department of Agriculture stressed research and education to help farmers improve their agricultural practices. In 1862, for example, the department began publishing reports that provided practical aid to

Old College Hall on the campus of Michigan Agricultural College (later Michigan State University), East Lansing, in 1862. Established in 1855, the institution was the first state college to offer instruction in scientific and practical agriculture and was the model for the Morrill Land Grant College Act.

farmers about the conditions of crops and the weather, acreage under cultivation, yield per acre, and the numbers of livestock raised on the nation's farms. Annual reports also presented a broad statistical survey of agricultural conditions and affairs in the United States.

In 1865, the department began agricultural investigations at an experimental farm. The farm was located in what is now the heart of Washington, D.C., below the Capitol on the Mall between 12th and 14th streets and Constitution and Independence avenues. (During the Civil War, the War Department had used the area as a cattle yard.) The Agriculture Department's scientists raised sorghum, wheat, rye, and vegetables on this plot of land.

In addition to initiating new activities, Newton pushed to improve the office space for the department. The offices for the new agency were housed in the

34

basement of the Patent Office. These quarters, however, were inadequate and Newton soon urged Congress to appropriate money for the construction of a new building. In 1867, Congress granted his request and provided $100,000 for a structure to be built on 14th Street and Independence Avenue in southwest Washington, D.C., and construction was completed a year later.

Commissioner Newton worked hard to help the department accomplish its goals. Unfortunately, his dedication to research contributed to his death on June 19, 1867. On a hot July day in 1866, Newton heard a thunderstorm approaching and worried that the storm might damage the wheat that the department's workers were harvesting on the experimental farm. As he hurried to the farm to ensure that all was well, he was overcome by the heat and suffered such serious sunstroke that his health was permanently broken.

During the late 1860s and early 1870s, the Department of Agriculture expanded its scientific research, particularly in the areas of plant and animal diseases. Scientists within the department experimented with chemicals to control or prevent insect damage to crops. Departmental officials also attempted to make the United States as self-sufficient as possible by growing experimental crops of tea and sugar beets in order to end reliance upon foreign

Isaac Newton (seated center), first commissioner of agriculture, and his staff in 1867. Under Newton's direction, the department emphasized research and education to assist farmers in improving agricultural practices.

The Department of Agriculture building in Washington, D.C., in 1869. The department's experimental farm, located just beyond the Washington Monument (left of center), was the site of many scientific investigations, including the growing of crops such as sugar beets and tea.

suppliers and to benefit farmers by giving them new crops to sell for a profit. Although the experiments to raise sugar beets ultimately proved successful, the efforts to raise tea were eventually abandoned as impractical because of the extensive amount of hand labor required to harvest the leaves.

In 1883, the department created a Veterinary Division to investigate animal diseases. After receiving congressional approval one year later, the division became the Bureau of Animal Industry. This expansion of the department enabled it to employ more scientists who had greater and more wide-ranging responsibilities. The creation of the Bureau of Animal Industry marked an important new direction for the Department of Agriculture—not only was it the first bureau established in the new department, but it also wielded regulatory powers that placed the Department of Agriculture in a much stronger position to aid farmers. Specifically, the Bureau of Animal Industry was authorized to enforce legislation passed in 1873 that provided for the humane treatment of livestock during shipment to market. It also helped the Treasury Department, which had responsibility for overseeing foreign trade, enforce regulations to prevent the importation of diseased livestock.

Although the department continued to seek new ways to improve agricultural production and farming practices, its regulatory activity during the 1880s was a major departure from its previous policies, which stressed research and education. Thereafter, the department slowly continued to expand its regulatory powers in order to aid agriculture by helping to eliminate crop and livestock diseases and unethical business practices, such as the selling of oleomargarine as butter.

During the 1880s, farmers and consumers increasingly became hostile to manufacturers of oleomargarine. Oleomargarine, a substitute for butter, was first produced by the meat packing companies, using the by-products of slaughtered cattle and hogs. Farmers objected to this competing, manufactured product because it could be sold more cheaply than butter and because manufacturers colored it yellow to make it look like butter. Farmers also charged that oleomargarine was unhealthy to eat. In 1883, as a result of this dispute, the Division of Chemistry, under Chief Chemist Harvey W. Wiley, began to examine food, particularly butter, for signs of adulteration. The division's chemists wanted to learn whether foreign substances had been added to the oleomargarine to change its purity and to determine whether the manufacturers were attempting to deceive the consumer by using words such as "butterine" on their product labels. These and later experiments compelled Congress to approve the Pure Food and Drugs Act of 1906, which regulated the processing and manufacturing of food products to protect the American public from contaminated and mislabeled foods.

In 1887, Congress supported the research activities of the department when it passed the Hatch Act (named after its sponsor, Representative William H. Hatch of Missouri). This statute authorized federal financial assistance to the states for the creation of agricultural experiment stations. Convinced of the stations' importance largely because of the valuable work performed by the experiment station of Wesleyan University in Middletown, Connecticut, Congress passed the act so that the land-grant colleges in each state could establish an experiment station, consisting of farmlands and laboratories where agricultural research would be conducted. Scientists would investigate new forms of technology, plant and animal diseases, the usefulness of commercial

The Union Stockyards in Chicago, Illinois, in 1900. With the expansion of railroads after the Civil War, western rangelands were opened up to the growing number of land-hungry farmers. Because of its rail connections with the East, Chicago became the meat-packing and shipping center of the country.

Dr. Harvey Wiley, chief chemist of the Division of Chemistry, led the campaign against impure foods during the 1880s. He and his staff investigated manufacturers of processed foods such as oleomargarine to determine whether they had contaminated or mislabeled products.

fertilizers, and the value of various plants used as livestock feed.

To coordinate the activities of the experiment stations and to ensure the uniformity of experimental procedures, the Hatch Act authorized the department to help the scientists at the experiment stations carry out their work. To fulfill this duty, in 1888 the department created an Office of Experiment Stations. This office served as a center for the exchange of scientific information to help the scientists at the various stations learn about important experiments elsewhere.

After the passage of the Hatch Act, the department began to provide both monetary and administrative assistance to cooperative federal-state agricultural projects. Cooperation at the state and federal levels helped Stephen M. Babcock, an agricultural scientist at the University of Wisconsin, to develop a testing device that measured the amount of butterfat in milk. Using this device, farmers could determine which cows produced milk with the most butterfat (milk with the most butterfat brought the highest prices), thus enabling them to keep their most productive cows and sell the others.

As the department's responsibilities expanded during the 1870s and 1880s, farm groups across the nation urged Congress to elevate the department to

The Poison Squad

On the morning of December 22, 1902, six young men sat down to a hearty breakfast at a table in the Bureau of Chemistry, located in the basement of the USDA building in Washington, D.C. They dined on oatmeal, eggs, potatoes, fruit, and bread. They did not know, however, that the butter they had eaten had been intentionally mixed with borax. Over the next five years, other teams of men ate nutritious meals of veal, roast beef, turkey, steak, and vegetables. Unknowingly, they consumed salicylic acid or benzoic acid and drank water mixed with sulfurous acid or milk contaminated with formaldehyde. These brave young men, collectively known as the Poison Squad, had volunteered to be human guinea pigs to help the Bureau of Chemistry determine the effects of chemical preservatives on the public's health.

Until 1876, foods were preserved by salting and smoking or by curing with vinegar or sugar. But, during the late 19th century, the food processing industry began using a multitude of chemicals to prevent spoilage and to improve the quality of foods. The bureau's chemists suspected that many of these chemicals were harmful. By 1900, it estimated that processors of food used 152 chemical preservatives and additives (chemical substances added to food to improve flavor, color, or texture). Peaches, for example, were uniformly colored yellow during the drying process. Most people did not know that they were eating a large

number of chemicals in their food—there was no federal law that required food processors to label products with their true ingredients. Furthermore, chemical preservatives were usually impossible to detect by smell or taste. Harvey W. Wiley, chief of the Bureau of Chemistry, was convinced that some chemical preservatives were hazardous to health, but he needed to prove his beliefs before he could persuade Congress to pass a law to regulate the use of chemicals in food. Wiley believed the best way to determine whether a particular chemical was harmful to one's health was to test it on people. He recruited young men who worked for the USDA, told them about the experiments he planned, and warned them about the risks. Each volunteer agreed to eat three meals a day at the bureau's "hygienic table" and refrain from consuming other foods during that day. Volunteers were instructed to keep records of their temperature, pulse rate, and how much they ate at each meal. The experiments usually lasted from 30 to 70 days. Each volunteer received a weekly physical exam to determine whether he was ill or incapable of continuing with the experiment.

All of the foods that the Poison Squad consumed were initially of the highest quality and free from commercial preservatives so that the chemists could carry out controlled experiments. Wiley chose borax for the first experiment because it was the most commonly used food pre-

Harvey W. Wiley (standing at the rear of the room), chief of the Bureau of Chemistry, oversees a Poison Squad meal.

servative. The chemical was mixed with butter, milk, coffee, and meat. However, members of the Poison Squad tried to guess which foods contained the chemical and avoided eating them. To counteract this tendency, Wiley had the volunteers swallow capsules of borax so that they would get the proper dosage needed for the experiment.

It did not take long for the volunteers to feel the effects. They suffered headaches, nausea, diarrhea, vomiting, constipation, back pain, and loss of weight and appetite. These results led Wiley and his chemists to conclude that borax was harmful to one's health and should not be used as a food preservative if safer chemicals were available.

Tests of other preservatives yielded similar results. For example, tests of formaldehyde (used for preserving milk) indicated that it produced rashes, caused insomnia, and irritated the liver and mucous membranes. The chemical also made the kidneys work extra hard to expel it from the body.

The experiments with the Poison Squad rapidly drew the attention of newspapers and magazines across the nation. People were shocked to learn of the use of preservatives in their food and demanded that Congress pass legislation to protect them from the dangerous chemicals.

President Theodore Roosevelt supported such regulations and urged passage of new food and drug laws. Finally, on June 30, 1906, Congress passed the Pure Food and Drugs Act, prohibiting the interstate sale of adulterated or mislabeled food. The law defined adulterated food as food that contained any substance that affected its quality or that had been substituted for an important ingredient claimed on the label—for example, the substitution of chicken for beef in a canned product. The law also prohibited food processors from using chemicals to conceal damaged or spoiled products, or from making false or misleading statements on the labels. Processors who violated the law could be fined, imprisoned, or both.

cabinet-level status to increase its prestige so that farmers would have a stronger voice in governmental affairs. As early as February 24, 1874, Congressman James Wilson of Iowa introduced a bill to grant cabinet-level status to the department. Although Congress did not approve his proposal, various agricultural groups, such as the Patrons of Husbandry, commonly known as the National Grange (first organized in 1867), which was dedicated to improving social and educational conditions for farmers, supported the movement to elevate the department to cabinet rank. During the next decade other bills were introduced to promote the department. Some congressmen and their constituents opposed these bills, however, because they believed that a Department of Agriculture within the executive branch would represent only farmers rather than the agricultural interests of all Americans. Other opponents believed that cabinet status would subject the secretary to partisan politics and that people would be hired to repay political favors rather than for their abilities.

Early in 1887, more than a dozen years after Congressman Wilson first introduced a bill to elevate the department to cabinet status, Congress finally approved the measure. Congress gave in to the public's support of the bill because the department had been relatively free from partisan politics. President Grover Cleveland signed the bill into law on February 9, 1889, and appointed Norman J. Colman to the new post of secretary of agriculture. Colman had been serving as commissioner of agriculture since April 1885. Although Colman took the oath of office on February 15, 1889, he had little opportunity to influence the direction of the department because President Cleveland had failed to win reelection and his administration came to an end in early March.

Jeremiah McLain Rusk was appointed the first full-term secretary of agriculture by President Benjamin Harrison. Secretary Rusk took office on March 6, 1889, and was determined to make the scientific research of the department more readily available to the farmers across the nation. To accomplish this, he introduced the publication of a series known as the *Farmers Bulletins*. In these regularly issued bulletins, farmers learned about the latest research and best techniques for raising crops and livestock. Each bulletin would be devoted to a particular topic, such as growing wheat, training horses, or building a barn. Indeed, throughout the late 19th century, the department tried to apply scientific research to practical farming in order to improve crop and livestock production.

One of the most important discoveries made by scientists in the Department of Agriculture during the 19th century occurred in 1889. Scientists in the

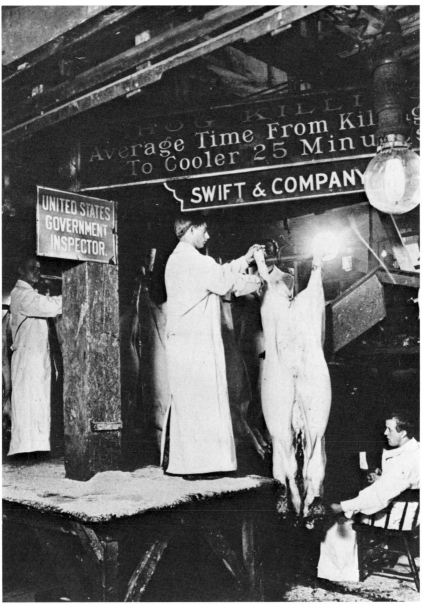

In 1906, inspectors examine slaughtered hogs for signs of tuberculosis at a meat-packing plant in Chicago. The Bureau of Animal Industry began to regulate the inspection of beef and pork in interstate commerce in 1890 to safeguard the health of consumers.

Bureau of Animal Industry found out that ticks carried organisms that caused cattle fever. The USDA's research enabled livestock producers to treat their cattle and to stop the disease from killing their herds. These scientists also proved that a disease could be transmitted by a carrier from one animal to another. The discovery had far-reaching effects because it helped scientists study whether other diseases could be spread by insects. Ultimately, the work of USDA scientists led to the discovery and prevention of other insect-related diseases that affected humans, such as yellow fever, malaria, typhus, and Rocky Mountain spotted fever.

During the 1890s, the department conducted other important work. The Bureau of Animal Industry, for example, helped to end the communicable disease pleuropneumonia (an often fatal respiratory disorder) in American cattle by 1892. The bureau also gained new responsibilities at this time, pushing the department toward a regulatory role. In 1890 and 1891, Congress passed meat inspection acts that authorized the department to inspect meat and live cattle and hogs shipped in interstate commerce to protect the health of the consumer. The department also worked to improve country roads so that farmers could easily take their produce to market. In addition, the USDA sought new and expanded markets in Europe for American farm products. And, in 1895, the department began the scientific study of nutrition.

In 1897, James M. Wilson became secretary of agriculture and a new phase in agricultural research was launched. His 16-year tenure is the longest served by any cabinet officer to date. Appointed by President William McKinley and reappointed by presidents Theodore Roosevelt and William Howard Taft, Wilson was affectionately known as "Tama Jim" after his home county of Tama, Iowa. Wilson had been a professor of agriculture, a director of the experiment station at Iowa Agricultural College, and a member of the Iowa State Legislature and the United States House of Representatives. Under his leadership, the USDA extended its scientific and regulatory responsibilities. Although Wilson was progressive and supported scientific reform in agriculture, he was a conservative in relation to nonagricultural affairs. For example, he rejected motion pictures, insisting that they were tools of the devil; he bought an automobile for the department in 1912, but only upon assuring his employees that he did not intend to purchase another one. The automobile was for use at an experimental farm in Beltsville, Maryland. (In 1910, the department bought farmland in Beltsville to carry out its research on animal husbandry and diseases. In 1932, it began moving its experimental activities from Arlington, Virginia, to Beltsville because of governmental growth—the War Department required the use of the space in Arlington for construction of

James M. Wilson, a former congressman from Iowa, was the fourth secretary of agriculture. He expanded the scientific and regulatory roles of the department from 1901 to 1904 by organizing the Bureau of Plant Industry, the Bureau of Chemistry, the Bureau of Forestry, and the Bureau of Entomology.

the Pentagon. The move of the experimental farm to Beltsville was completed in 1942.)

In 1901, Wilson organized the Bureau of Plant Industry to consolidate ongoing research concerning plant life. He also sent "plant explorers" to distant parts of the world to find new species that might be suitable for American agriculture. These explorers brought back a short-grain Japanese rice that boosted production in the South. They also collected Kharkov and Kubanka wheat from Russia, wheats that were hardy enough to withstand the droughts and frosts of the Midwest. These varieties helped improve the cultivation of hard red winter

45

One of the Bureau of Plant Industry's explorers collects samples during an expedition on the island of Sumatra, Indonesia, in 1926. Bureau explorers traveled to remote areas of the world to gather new plant species that might be applicable to American agriculture.

wheat in the northern Great Plains. Other plant introductions included Peruvian alfalfa, Venetian sweet peppers, Egyptian cotton, and ornamental shrubs such as the Chinese Elm and flowers such as the Meyer Rose. Scientists in the bureau used these foreign plants to develop new varieties that would resist disease and pests and that would adapt to a variety of growing conditions—all for the purpose of helping farmers become more productive and profitable.

In 1901, Secretary Wilson also created the Bureau of Chemistry. The bureau's chemists examined manufactured, or processed, foods to determine whether they contained harmful chemical preservatives and whether the contents of the foods were what the manufacturers claimed in advertisements and on the labels. The chemists had good reasons to be suspicious of the food manufacturers. A popular jingle of the time clearly demonstrates the prevailing view of the corruption that existed in the food industry: "Mary had a little lamb, and when she saw it sicken, she shipped it off to packing town and now it's labeled chicken."

The pioneering work of the Bureau of Chemistry and the publication in 1905 of Upton Sinclair's novel *The Jungle*, which revealed the atrocious and insanitary working conditions in the slaughter and meat-packing houses of Chicago, convinced Congress to pass the Pure Food and Drugs Act and the Meat Inspection Act in 1906. The new food law required manufacturers to label their products accurately and to refrain from adding prohibited chemicals to improve the color, flavor, or preservation of the foods. The Meat Inspection Act, administered by the Bureau of Animal Industry, enabled the department's inspectors to help ensure that slaughtering conditions were sanitary and that diseased livestock were not butchered and sold to the consumer.

Another major achievement for Wilson was the creation of the Bureau of Forestry in 1901. Four years later, the forest lands administered by the General Land Office of the Department of the Interior were transferred to the Bureau of Forestry. Wilson established the Forest Service in 1905 to administer and protect the nation's forests. He ordered his chief forester, Gifford Pinchot, to safeguard the nation's forests to preserve them for the enjoyment of all the people and to protect them from exploitation by individuals and companies. In 1907, the nation's forest reserves were renamed national forests.

In 1904, Secretary Wilson formed the Bureau of Entomology to study plant diseases. By the time Wilson left office in 1913, 35 laboratories across the nation conducted research to eliminate pests, such as the Mexican boll weevil, gypsy moth, and San Jose scale, that devastated crops. The research of the scientists in the Bureau of Entomology led to passage of the Plant Quarantine

A boll weevil on a young cotton plant. The department sent extension agents into the southern states to help cotton farmers eliminate this crop-destroying pest.

Act in 1912, which regulated the importation and shipment of foreign plants to prevent the introduction or spread of plant diseases and pests.

During Wilson's tenure as secretary of agriculture, the department's first agents, under the direction of Seaman A. Knapp, began work in 1904. Their demonstration work led to the passage of the Smith-Lever Act in 1914, which created an official Extension Service, cooperatively administered by the Department of Agriculture, the state agricultural colleges, and local governments. The agents were to be joint employees of the USDA and the agricultural colleges, but the colleges or local governments paid their salaries, in some cases from funds allocated by the department. The Extension Service quickly became responsible for making farmers aware of the most useful scientific developments that would help improve their agricultural operations—

including fighting insects and crop diseases and learning new methods for planting crops and conserving the soil.

Around the time of the passage of the Smith-Lever Act, a group of USDA agents was sent into the South to teach cotton farmers how to combat the boll weevil. The boll weevil is a beetle with a long snout that punctures a hole in the young cotton boll and then lays its eggs there. The larvae feed on the maturing cotton, thus destroying the crop. Farmers fight the boll weevil by using insecticides, early maturing varieties of cotton, and good cultivation practices in their fields, although they have not been able to eliminate this pest entirely.

On August 1, 1914, as the extension agents began work in the field, Europe became embroiled in a major war. During the Great War, or World War I, the USDA encouraged farmers to increase production on their farms, particularly after the United States entered the conflict in 1917. Demand for great quantities of food rose because there were fewer farmers producing crops—

American troops in combat during World War I. The wartime demand for food and the higher prices for commodities prompted the USDA to urge farmers to increase agricultural production.

farmers were needed as soldiers—and those American farmers not in the military service responded by increasing their output.

During the war, the USDA worked closely with the Food Administration, which had the responsibility for organizing the distribution and conservation of food. Both agencies cooperated in food conservation campaigns. For example, on January 26, 1918, Herbert Hoover, the brilliant administrator of the Food Administration, asked the American people to observe wheatless Mondays and Wednesdays, meatless Tuesdays, porkless Thursdays and Saturdays, and the use of victory bread (a bread with corn flour or oatmeal used to replace wheat flour because wheat bread was needed for the troops in the war).

Because of wartime demands, the prices of agricultural commodities rose sharply and remained high until 1920, when the economic recovery of the European nations decreased foreign reliance upon American farm products. As demand diminished, the prices of agricultural products dropped as well. Soon, farmers were producing more grain, cotton, and livestock than they could ever hope to sell.

As agricultural prices plunged, industrial prosperity continued to accelerate. Many farmers believed that they were not being treated fairly in the marketplace. Although some seemed to prosper, more and more farmers lost money. Soon, they began to advocate government aid to agriculture. They wanted the federal government to take an active role to open new markets, to guarantee prices at certain levels, and to require that producers of surplus commodities reduce production.

Henry C. Wallace, secretary of agriculture under President Warren G. Harding, sympathized with the farmers' demands. He believed that the USDA was "charged with the duty of promoting agriculture in its broadest sense." Throughout the 1920s, however, Presidents Warren G. Harding, Calvin Coolidge, and Herbert Hoover would not permit the USDA to embark upon a new course of action that involved helping farmers market their crops or providing monetary relief for those on the verge of bankruptcy. Nevertheless, Presidents Coolidge and Hoover did support federal aid to help farmers organize cooperatives to buy and sell collectively. Coolidge and Hoover believed that cooperatives could withhold low-priced commodities, such as wheat and cotton, until prices increased and that the unity of action would enable farmers to sell their products for the highest possible price and buy needed items at the lowest possible cost.

The government's hands-off policy in dealing with agriculture made a bad economic situation worse for the American farmer. When the economy collapsed after the stock market crash in 1929, farmers nationwide believed

Secretary Henry C. Wallace was unable to persuade President Harding to allow the department to help farmers market their crops or to give financial assistance to those who were on the brink of bankruptcy.

that the time had come for direct federal intervention. During the first 70 years of the Department of Agriculture's existence, its scientists had helped farmers vastly increase production through research and education. But during the Great Depression, many farmers were poverty stricken, in part because they were too productive. The time had come for the department to create a new plan that would offer the farmers relief.

Reform was forthcoming. When President Franklin D. Roosevelt took office on March 4, 1933, he reiterated what many farmers had already said: "This nation needs action and action now." The economic situation on America's

A 1932 cartoon depicts Franklin D. Roosevelt assisting a crippled old man, who represents the American farmer during the Great Depression. After Roosevelt became president in 1933, he championed several federal programs to help solve the farmers' problems of low prices and surplus production.

farms was so grave that the president of the American Farm Bureau Federation, the largest and most conservative farmers' organization in the nation, warned, "Unless something is done for the American farmer we will have revolution in the countryside within less than 12 months." The agricultural situation was indeed serious—farm income had fallen by one-third since 1929 and agricultural prices had decreased more than 50 percent.

President Roosevelt unquestionably understood the severity of the agricultural crisis. To help solve the farmers' problems of low prices and surplus production, he appointed Henry A. Wallace as secretary of agriculture. Henry A. Wallace was the son of Henry C. Wallace, who had served as secretary of agriculture under President Harding. Secretary Wallace believed that the solution to the agricultural crisis, in part, was to establish guaranteed prices for produce. USDA officials quickly drafted a bill for congressional approval.

Congress passed the legislation, known as the Agricultural Adjustment Act, and President Roosevelt signed it into law on May 12, 1933. The act broadened the powers of the USDA by authorizing it, through the Agricultural Adjustment Administration (AAA), to pay farmers to reduce production of certain crops, such as wheat, cotton, corn, rice, tobacco, pork, and milk—commodities that were in large supply. The department soon expanded the list of commodities to include rye, flax, barley, grain sorghum, sugar beets, sugarcane, potatoes, and peanuts. After 1933, the secretary of agriculture announced price supports for specific crops before each planting season, so that farmers could plan whether to participate in the program before they sowed their crops. The prices were based on national averages determined over a period of several years. The funds for these payments were to come from a special tax levied on processed food manufacturers.

USDA officials believed that if production of surplus commodities could be reduced, the supply would decrease and the prices paid for those goods would increase. In the meantime, the federal government would pay the farmers to reduce production and keep certain lands idle, and AAA funds would help keep farmers on the land until better times returned.

Plans for farmers to plow under, or bury, crops, such as wheat and cotton, and to slaughter hogs to reduce surplus production presented great dilemmas. The federal government found itself in a quandary: How was it going to fight hunger when it was asking for a reduction in crop surpluses at the same time? With tens of thousands of people suffering severe poverty and hunger during the Great Depression, many people were angry that the USDA intended to pay farmers to destroy crops and livestock that could be used to feed the nation's poor and to plow under cotton that could be used to clothe them. Secretary

Wallace recognized the problem, but he explained: "To have to destroy a growing crop is a shocking commentary on civilization. I could tolerate it only as a cleaning up of the wreckage of the old days of unbalanced production."

The AAA slowly reduced the commodity surpluses and its checks kept farmers on the land, gradually improving the economy and increasing prices. From 1932 to 1935, farm income rose 50 percent. But in 1936, the United States Supreme Court ruled that the Agricultural Adjustment Act was unconstitutional on the ground that the tax on the processed food manufacturers enabled the federal government to control agricultural production—an unconstitutional invasion of states' rights.

Although the Supreme Court struck down the Agricultural Adjustment Act, the nation's farmers had supported the legislation. Edward O'Neal, president of the American Farm Bureau Federation, called all those who opposed the AAA "enemies of the Republic." He promised, "There will be neither surrender nor compromise, as we move forward. . . . The principle of farm adjustment, in terms of supply and demand is not dead. . . . In fact, only the legal clothes of farm adjustment have been declared unsuitable."

Secretary Wallace agreed with O'Neal. Quickly, USDA officials drafted new legislation for congressional approval that included a crop reduction and price support program that would meet the constitutional test set by the Supreme Court. The USDA's new program enabled it to pay farmers for withdrawing acreage from production in order to plant soil-conserving grasses and legumes, such as alfalfa. The department could then continue its efforts to reduce surplus production, provide needed income for farmers until prices rose, and support soil conservation.

Congress approved this legislation, known as the Soil Conservation and Domestic Allotment Act, on February 29, 1936. Two years later, the program was expanded when Congress approved the second Agricultural Adjustment Act. In addition to supporting the production of soil-conserving crops instead of those that contributed to price-depressing surpluses, the Agricultural Adjustment Act of 1938 enabled the USDA to regulate the marketing of crops to help increase their prices. Under this AAA program, the department also provided loans to farmers to help them keep their crops out of the market until prices increased.

During the 1930s, Congress approved other programs that enabled the department to loan money to farmers for the purchase of seed, equipment, livestock, gasoline, and oil in order to maintain their farms, through agencies such as the Resettlement Administration, Farm Security Administration, and Soil Conservation Service. These loans helped farmers in the dust bowl (the

An abandoned farmstead in Oklahoma, photographed by the Soil Conservation Service in 1937, illustrates the devastating effects of wind erosion. During the mid-1930s, farmers in the western states suffered from the consequences of severe drought and dust storms. The Soil Conservation and Domestic Allotment Act of 1936 enabled the USDA to pay farmers to plant soil-conserving plants such as alfalfa.

region of the prairie states that is susceptible to dust storms) to fight drought, soil erosion, and economic depression. By the late 1930s, the USDA administered wide-ranging programs that provided for acreage reductions, marketing quotas, loans, and parity payments. (Parity payments were given to farmers when the prices of their commodities fell below a certain mark in order to help them maintain a standard of living comparable to that of Americans who worked in areas other than agriculture.) These programs were designed to help raise agricultural prices, increase farm income, and improve the quality of life in rural America.

During the Great Depression, the Department of Agriculture had vastly increased its authority and responsibilities. It no longer merely supported research, education, and regulations to ensure the public health. By the late 1930s, it had become an economic action agency that provided direct monetary assistance to farmers and regulated the production and sale of their produce. New agencies under the department, such as the Soil Conservation Service, supplied economic and technical aid to farmers to help them conserve the soil and improve their operations. Never before had the USDA become so involved with the daily activities and lives of the nation's farmers. Some people, of course, did not believe that it had devised the best policies, but those farmers

Secretary Henry A. Wallace addresses a group of Vermont and New Hampshire farmers in 1937. Wallace helped to convince Congress to pass the second Agricultural Adjustment Act, which enabled the USDA to regulate crop marketing and provide loans to farmers.

In 1940, a Maryland farmer fills out a crop report for the Division of Agricultural Statistics to help the department obtain information on farm conditions for its marketing service.

who had suffered severe economic hardship were great supporters of the department's programs.

Not all of the economic problems that farmers faced had been solved by the end of the 1930s. The department's programs to aid farmers had been beneficial—prices increased, surpluses declined, and many farmers avoided bankruptcy and remained on their land. When World War II began in Europe in 1939, demands for food and fiber, such as cotton and wool, increased dramatically and the prices of farm commodities rose. By the time the United States entered the war on December 8, 1941, the nation's agricultural economy was on the way to recovery from the cataclysmic years of the Great Depression. With that recovery, the Department of Agriculture would have new concerns and responsibilities, but also a new challenge: How would it cope with the tremendous increases in agricultural production made by smaller numbers of farmers?

During World War II, American soldiers in Italy feed a child some of their army rations. USDA scientists refined methods used to dehydrate and preserve meats and vegetables in order to make it easier for American troops to carry foods in battle.

FOUR

Years of
Readjustment

W orld War II created a great demand for food, livestock, and fiber produced by American farmers. As demand increased, prices rose significantly. Soon surplus production and low prices were problems of the past. The USDA urged farmers to expand production quickly in order to take advantage of the high wartime prices. But in January 1942, Congress passed legislation limiting price increases for crops and livestock, because food bills were rising and consumers were complaining. The Emergency Price Control Act placed a ceiling on farm prices—that is, an upper limit beyond which prices could not increase. And it was the USDA's responsibility to enforce the federal government's price control policy.

During the war, the department's scientists experimented with new methods for preserving foods and for reducing their weight and bulk in order to make them more usable by the American troops in combat. These scientists perfected a way to dehydrate meats and vegetables and to combine them in prepackaged soups and stews. They learned how to dry milk and eggs and conducted research on freezing foods, which helped to expand the frozen-food industry after the war. USDA scientists also experimented with new chemical pesticides to help protect the soldiers, who often had to live in filthy conditions, where insects teemed. After the war, these chemical pesticides, such as DDT (which today is prohibited in the United States), were made available to farmers to help them eradicate insects and other pests from their fields, orchards, and livestock.

In 1942, President Franklin D. Roosevelt expanded the authority of the secretary of agriculture to help the department contribute to the war effort. The secretary's increased responsibilities were (1) to determine the food requirements at home and abroad for both civilians and the military; (2) to develop and execute a program designed to meet all of those food needs; (3) to assign food priorities and make allocations, that is, require rationing; (4) to ensure the efficient distribution of agricultural commodities to the nation's markets; and (5) to formulate policies to govern the purchase of food by federal agencies for export or relief purposes or domestic use.

In addition to these responsibilities, the department helped the nation's farmers increase food production each year. The USDA's research on nutrition also marked one of the most important long-term gains from the war and helped the federal government support a nationwide campaign to improve the diet of all Americans. Furthermore, USDA officials worked with the allies of the United States to help feed people whose lives had been disrupted or their farms ruined by the war. Wartime cooperation between the USDA and the British Food Mission led to an international conference in 1943 in Hot Springs, Virginia. This conference resulted in the creation of the Food and Agriculture Organization of the United Nations after the war.

When the fighting ended in 1945, famine threatened some regions of the world because the war had destroyed many farms and disrupted agricultural trade. During the three years following the victory in Europe in 1945, the department supplied food for inhabitants of the war-torn nations of the world. Between July 1, 1946, and June 30, 1947, the department organized the shipment of 400 million bushels of wheat and 168 million bushels of other grains to people abroad who faced hunger and starvation.

By the late 1940s, many of the war-torn nations were recovering. The farms of France and Germany, for example, were productive again and, as a result, the demand for American agricultural products diminished. As demand declined, agricultural prices fell in the United States. Many farmers worried that they were about to experience another economic depression like the one that followed World War I. When prices began to fall, the USDA administered a price support program, authorized by Congress, that kept the prices of agricultural commodities high enough so that farmers could earn a profit. The department purchased large quantities of wheat, corn, and dairy products, thus removing surpluses from commercial markets and keeping demand and prices as high as possible.

In 1959, a Greek woman and child prepare food received in a CARE (Cooperative for American Relief to Everywhere) package, which contained powdered milk, flour, cornmeal, and other foods from USDA surplus-food warehouses.

Before the USDA could resolve the problem of high production and declining prices, the United States entered the Korean War in 1950. Once again, wartime demands encouraged farmers to expand production to take advantage of high prices. When the Korean War ended in 1953, farm surpluses had again accumulated. Science and technology—new implements, machinery, seeds, fertilizers, and chemicals—enabled farmers to be more efficient than they had ever been before. To dispose of the agricultural surpluses and to keep prices high enough to support an adequate standard of living for farmers, the department established marketing quotas, or limits on the amount of crops farmers could sell on the market. These quotas were applied to wheat, cotton, tobacco, and peanuts. The USDA also continued its price support program by granting loans to farmers for major commodities, such as wheat, corn, and cotton, and sought ways to expand foreign trade.

The enactment of the Agricultural Trade Development and Assistance Act of July 10, 1954, enabled the department to develop new markets and reduce surpluses. Commonly known as Public Law 480, this legislation allowed the Department of Agriculture to sell farm products for foreign currency—German marks, French francs, and British pounds—rather than requiring payment in American dollars. (This procedure enabled foreign nations to buy American agricultural products whenever those commodities were needed. They did not need to wait until they sold something to the United States in order to have American currency to make their agricultural purchases.) It also authorized the department to ship food abroad for relief purposes and to trade the farm products owned by one of its agencies, the Commodity Credit Corporation, for materials that the federal government needed. Under Public Law 480, the department emphasized foreign trade and the sale of agricultural commodities abroad.

Despite Public Law 480, the USDA was unable to reduce sufficiently the supply of agricultural commodities that it had acquired from farmers through the price-support loan program. Although the department tried to control much of the surplus production of cereal grains and dairy products in order to keep prices high in the marketplace, farmers became increasingly efficient and produced large harvests each year. To remove agricultural surpluses from the market when prices fell, the government had to spend more money. In 1952, for example, the federal government, through the USDA, spent $288.6 million to control surpluses and to stabilize prices. By 1959, the amount skyrocketed to $2 billion. Storage costs for the surpluses also increased rapidly. In 1952 the federal government spent $73.2 million to store and handle surplus agricultural

A farm at dusk. The Rural Electrification Administration gives loans to rural residents who join cooperatives to obtain electricity, enabling farmers to use power tools and machinery.

commodities. In 1959 the costs had increased more than sixfold to $481.6 million.

Because farmers continued to increase production beyond the department's ability to purchase surpluses, Congress passed the Agricultural Act of 1956. This legislation established a Soil Bank Program, designed to decrease the agricultural surpluses by removing land from cultivation. This would reduce the production of price-depressing crops. The Department of Agriculture paid farmers to divert acreage planted in surplus crops, such as wheat, cotton, tobacco, peanuts, and rice, to other crops. Or, farmers could let all or a portion of their land lay idle and allow it to return to grassland. In return for not planting any crops at all on certain land, the government paid farmers a fee.

Although nearly 29 million acres of land had been withdrawn from production by 1960, the surplus of farm commodities did not lessen in quantity because farmers became even more productive on fewer acres by using new discoveries in science and technology, such as hybrid corn (a breed of corn obtained when two different varieties are crossbred), chemical fertilizers, and pesticides and herbicides. By the late 1950s, the USDA had great difficulty persuading farmers to reduce farm production even by paying them to do so. It also had engaged in an expensive price-support program—the department acquired agricultural commodities for federal storage facilities in order to keep large quantities of surplus crops out of the commercial market, thereby keeping prices as high as possible to benefit farmers.

During the 1960s, the Department of Agriculture attempted to manage the problem of surplus production and low prices by limiting the amount of wheat and cotton that farmers who agreed to take part in the program could sell and by imposing acreage allotments, or the maximum amount of land that could be planted with certain crops. The department encouraged farmers to raise crops that were not overabundant. The USDA continued to support agricultural prices through the loan program of its Commodity Credit Corporation (CCC). (The CCC was established as an agency of the Reconstruction Finance Corporation, a federally sponsored loan organization, in 1933 to help improve farm income and to encourage the effective and orderly marketing of agricultural commodities. It was transferred to the USDA in 1939.) It bought, sold, and exported agricultural products and made loans to farmers. Farmers used the loans to pay for daily operating and living expenses while they waited for agricultural prices to rise. If prices did not increase enough to pay off the loans and to earn the farmers a profit, they could give their crops to the department as repayment of the loans.

The Rural Electrification Administration (REA) also conducted some of the department's most important work. Created in 1935 and transferred to the USDA in 1939, the REA made loans to electric cooperatives in order to provide electric service to homes and farms that were without it. Primitive conditions abounded on most of the nation's 7 million farms during the 1930s—three-fourths of the rural population had no indoor plumbing and 90 percent lacked electricity. By 1970, 98 percent of the country's farms had electricity, improving the farmers' quality of life and enabling them to use a host of electrically powered tools and equipment, from saws and cream separators to washing machines and hot-water heaters. The REA furnished the expertise, but it depended on the farm community's enthusiasm to form cooperatives. For instance, farmers could join a cooperative for a $5 fee and qualify for loans to install power lines; the loans could then be paid off over a 25-year period. The REA also supported the development of industries in agrarian areas by providing technical expertise and by helping developers locate needed financing. Projects supported by the USDA provided new jobs and improved the standard of living in rural areas.

The department also continued to ship food supplies to nations in need under Public Law 480, now commonly known as the Food for Peace program. India, Pakistan, Yugoslavia, Brazil, Spain, the United Arab Republic, and the Republic of Korea received millions of dollars worth of food supplies under this program. By 1970, the Food for Peace program had helped feed 47 million children in Brazil, Sierra Leone, India, Korea, Jamaica, Chile, and the Philippines. Although the Department of Agriculture provided food shipments from its warehouses that stored commodities acquired under various price support programs, religious and other relief agencies distributed the food reserves.

A Missouri farmer feeds his cattle. The USDA provides special livestock loans through the Farmers Home Administration that can be used to buy feed; to restock a depleted herd; and to pay for the maintenance of a ranch.

In Calcutta, India, dockworkers load bags of wheat onto a truck. Through Public Law 480 and the Food for Peace program, the USDA provides surplus commodities, such as wheat, to foreign countries in need.

Also by 1970, more than $19.6 billion worth of agricultural commodities had been shipped abroad to nations in need since the program began in 1954. Throughout the decade, the USDA continued to help developing nations improve agricultural production, to decrease hunger and starvation, and to improve nutrition, health, and standards of living, all the while reducing agricultural surpluses in the federal government's warehouses and granaries.

The department was also increasingly emphasizing foreign trade by seeking new markets, expanding old ones, and offering technical assistance to other nations. For example, with USDA assistance Brazil developed agricultural cooperatives, improved its agricultural credit system, and studied the soil in its undeveloped interior. In Kenya, USDA researchers were able to develop hybrid varieties of corn that produced large crops in areas of high rainfall and other varieties that produced abundantly in regions with average or below average annual precipitation. In other parts of Africa, USDA scientists helped nations develop new and more productive varieties of millet (a grass cultivated for its grain), sorghum, and corn. The department also brought foreign agricultural scientists to the United States so that they could advance their studies. In 1970 alone, more than 3,600 foreign agricultural scientists and technicians from 128 countries studied advanced farming techniques in the United States.

World food shortages and greater foreign purchasing power increased the demand for American agricultural commodities. Exports expanded and surplus reserves were quickly diminished. Large sales of grain to the Soviet Union in 1972 and 1973 helped boost the shipment of agricultural commodities abroad by

Harvesting corn from a field that has been tilled using soil conservation methods; after the corn has been harvested, the parts of the stalks remaining in the soil will be left there. Later, the field will be chiseled to open up the subsoil for future planting.

25 percent. As the reserves of agricultural commodities declined, food prices increased, and consumers advocated ending department programs designed to reduce production in order to increase prices. By 1973, consumers spent nearly $1.42 for food that had cost only $1 in 1967. Many consumers blamed the department's policies for these price increases.

The inflation of food prices, together with increased exports, made many Americans more aware of USDA policy-making. No longer did many view the department as merely an agency that supported science, education, and health regulations. The public recognized that the department influenced the food budgets of every household and the pocketbooks of all Americans.

The department responded to these new economic conditions and public concerns by urging farmers to expand production and by limiting price supports for grain and cotton. Secretary of Agriculture Earl L. Butz, whose tenure began in December 1971 and ended in October 1978, recognized this change

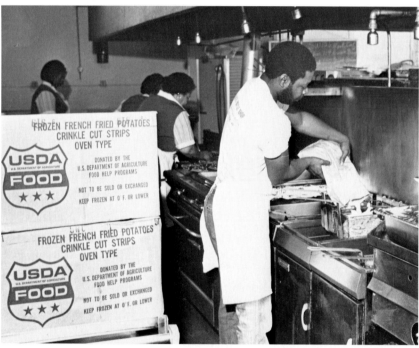

Cooks in Oklahoma City, Oklahoma, prepare food donated by the USDA. The department's Food and Nutrition Service makes surplus goods available to children in school lunch and breakfast programs and to other people who need public assistance.

in department policy as "an historic turning point in the philosophy of farm programs in the United States." In contrast to earlier programs dating from the New Deal that attempted to limit production, during the mid-1970s the department urged farmers to raise more wheat, corn, upland cotton, and tobacco. Increased production, however, caused prices to drop. Between 1974 and 1977, the price of wheat decreased from $4.09 to $2.33 per bushel and the price of corn dropped from $3.02 to $2.02 per bushel. Moreover, net farm income (income remaining after all operating expenses have been deducted) fell 42 percent. The department responded by returning to an earlier policy of reducing production by establishing acreage allotments—the amount of land farmers could plant and still be eligible for price supports on certain crops.

During the 1980s, under Secretary John R. Block, the USDA attempted to reduce the government's involvement in agricultural price support and loan programs by relying more on exports. Because agricultural groups such as the Farm Bureau advocated additional aid to agriculture, however, Congress did not substantially change the responsibilities of the department when it passed the Agriculture and Food Act of 1981. Support prices (the prices that the federal government loaned farmers on their commodities) and acreage allot-ments remained unchanged. Nonetheless, in an effort to cut costs, Congress reduced the amount that farmers could receive in payments from the federal government. No farmer could receive more than $50,000 annually for loans and no more than $100,000 for disaster relief (payments for loss of crops to natural disasters such as floods, hail, and wind). The law also increased the value of food products that could be donated to other countries to $1 billion.

Because of a worldwide recession that reduced demand for American agricultural products and because of a decreased foreign purchasing power, the Agriculture and Food Act of 1981 did little to help the department resolve the problems of overproduction and low prices. Exports declined and prices dropped. Many farmers went bankrupt and gave up their farmland. Farmers who had taken government money in price support loans to help them hold their crops off the market while waiting for higher prices soon found that better prices were not forthcoming. As a result, many defaulted on their loan payments and had to turn over their crops to the federal government.

To combat the annual problems of surplus production and low prices, the USDA returned to an acreage reduction program for wheat, cotton, rice, and feed grains. To be eligible for price supports, farmers had to reduce their planted acreage of wheat, cotton, and rice by 15 percent and their acreage of feed grains by 10 percent. If, for example, a farmer normally planted 100 acres in wheat, he could plant only 85 acres to remain eligible for the government's

Harvesting wheat in Pullman, Washington. To combat overproduction, the USDA established the Payment-In-Kind program, which pays farmers not to grow certain crops in order to limit production and diminish government-owned stocks of commodities. Payments are made in the form of goods from the Commodity Credit Corporation's storage facilities.

price support program. Because of improved fertilizers, pesticides, and machinery, however, farmers produced more of these crops on fewer acres than ever before. Consequently, the department failed to reduce surplus production.

To fight overproduction, Secretary Block announced a new farm program in January 1983. Called the Payment-In-Kind (PIK) program, it was designed to limit production and to reduce the stocks of government-owned agricultural commodities by paying farmers not to raise certain crops. These payments, however, would be made in the form of commodities from government storage facilities (the Commodity Credit Corporation's warehouses). The farmers then could sell these commodities. This program covered wheat, rice, upland cotton, corn, and grain sorghum. To participate farmers had to halt cultivation of these crops on between 10 and 30 percent of their lands. The federal government paid them at a rate of 80 percent of normal yields for all crops except wheat, for which they paid at a rate of 90 percent. If an acre of land, for example, produced 100 bushels of corn, the government agreed to pay the farmer 80 bushels of corn per acre not to plant that crop. The farmer gained a riskfree payment and avoided operating expenses by supporting this program. By paying farmers in kind, the USDA reduced the reserves of the CCC and enabled farmers to limit production and yet earn a satisfactory income. (However, these farmers could not take part in other USDA loan or price support programs.)

Many farmers participated in the PIK program. Soon 82 million acres were withdrawn from production and more than one-third of the land normally planted with crops was taken out of production. High participation, however, meant substantial costs for the federal government. In 1983, the USDA paid farmers more than $9 billion in commodities to participate in the program.

By the late 1980s, the price support programs had changed relatively little

Excess wheat at a grain elevator in Creston, Washington. The USDA works with farmers to resolve the problems of overproduction by instituting acreage-reduction programs and by expanding exports to foreign nations. For example, the USDA sold grain to the Soviet Union in 1971 and 1972.

A USDA entomologist examines a tomato plant for symptoms of the potato spindle tuber viroid at the experimental farm in Beltsville, Maryland. Scientists at the farm conduct research in genetic engineering in order to produce disease-resistant plants.

since the policy was first introduced during the 1930s. The department still used price supports to help solve the perennial problem in American agriculture—the ability of farmers to produce more than the public could consume at home and the department could export abroad. Although the USDA worked to control surplus production and to ensure a fair income for farmers, changes in science and technology continued to make farmers more productive on fewer acres of land. As the department relied more and more on exports to reduce surpluses, prices became more uncertain because of uncontrollable events, such as social and political unrest abroad and bad weather. Prolonged drought in Africa, for example, increased demands for agricultural products under Public Law 480 (the Food for Peace program).

During the late 1980s, the USDA grappled with surplus production, low prices, and inadequate foreign markets and at the same time maintained its more traditional responsibilities in the areas of research, education, and regulation. Scientists at the agricultural experiment stations and at the department's experimental farm at Beltsville, Maryland, conduct research in a multitude of areas, including genetic engineering designed to produce disease-resistant and higher-yielding plants and livestock with more meat and less fat or more wool. County agents still inform farmers about ways to improve agricultural practices in their fields, workshops, barns, and homes based on information gained from practical research at the state agricultural colleges, the experiment stations, and in the USDA's laboratories and experimental fields. USDA employees regulate the processing and marketing of meats and poultry at packing plants and guard against the importation of diseased or pest-infected plants and livestock.

The department fulfills these responsibilities and tries to balance supply and demand in order to increase farm income while keeping government expenditures under control.

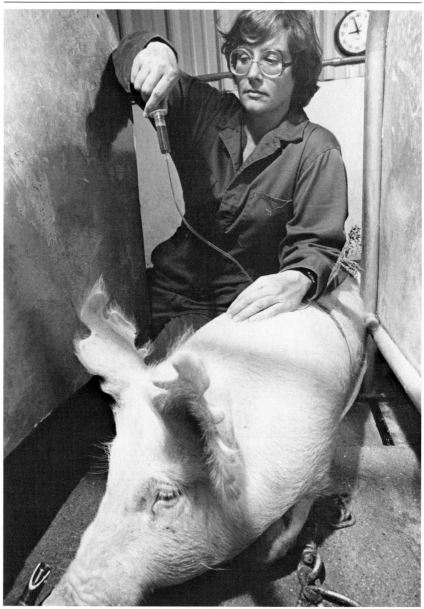

A USDA environmental physiologist takes a blood sample from a confined hog in an animal stress study at the U.S. Meat Animal Research Center in Clay Center, Nebraska. The study will determine what effects modern methods of raising livestock have on farm animals.

FIVE

The USDA Today

Although the USDA has undergone many structural transformations since Commissioner Isaac Newton took office to promote aid to farmers more than 125 years ago, its principal functions have endured. The USDA continues to work to improve and preserve farm income and to expand markets abroad for American agricultural commodities. It strives to alleviate hunger, poverty, and malnutrition. It assists landowners in protecting their water, soil, forests, animals, and plants in order to safeguard these resources for future generations. The USDA accomplishes these responsibilities through the endeavors of its more than 112,600 employees.

Executive Administration

Leading the department in its numerous duties is the secretary of agriculture, one of the 13 members of the president's cabinet. The secretary is appointed by the president to serve as his chief adviser on agricultural matters. The

The USDA's headquarters at 14th Street and Independence Avenue in southwest Washington, D.C.

secretary is supported in his or her job by a deputy secretary of agriculture who serves as the secretary's second-in-command and takes over the management of the department when the secretary travels to meetings.

The secretary and deputy secretary are assisted in the administration of the department by two undersecretaries, four offices offering administrative support, and seven assistant secretaries. The undersecretary for international affairs and community programs supervises the Agricultural Stabilization and Conservation Service, the Foreign Agricultural Service, and the Office of International Cooperation and Development. The undersecretary for small community and rural development manages the Farmers Home Administration, the Federal Crop Insurance Corporation, and the Rural Electrification Administration.

Barley at a grain elevator in Harrington, Washington, is loaded for shipment. The Commodity Credit Corporation works to stabilize food prices by making commodity loans to farmers; by purchasing produce; and by paying farmers to reduce planted acreage.

International Affairs and Commodity Programs

The Agricultural Stabilization and Conservation Service (ASCS) manages commodity and land use programs designed to reduce production in order to help stabilize prices, markets, and farm income. The ASCS administers the stabilization programs of the Commodity Credit Corporation (CCC) by making commodity loans to farmers, by purchasing their produce, and by paying producers to reduce their planted acreage or to limit the marketing of particular crops. This agency, managed by a seven-member board (appointed by the president and supervised by the secretary of agriculture), also works to conserve and improve soil and water resources and to protect cropland from erosion.

The CCC helps to stabilize, support, and protect farm prices and incomes and maintains adequate supplies of agricultural products. The commodities that the CCC acquires are disposed of by the USDA (through the commodity office located in Kansas City, Missouri) by means of foreign and domestic trade and transfers to other governmental agencies or through donations to needy people in the United States.

The Foreign Agricultural Service (FAS) reports on agricultural conditions worldwide. FAS employees serve at approximately 100 embassies and consulates around the world, studying trade, research, market development, and export programs in more than 110 countries. The FAS also helps to develop new markets and to expand old ones. It is responsible for sending government-owned (CCC) commodities to foreign nations, private companies, and nonprofit organizations (trade fairs and state agriculture departments, for example).

The Office of International Cooperation and Development (OICD) provides technical assistance and agricultural training to other nations, particularly developing countries, forming closer ties between the United States and the other nations. Officials in this agency work to solve international food problems and sponsor the exchange of scientists and information gained from research to help farmers both in the United States and overseas. Although the Department of State has the chief responsibility for foreign affairs, it often relies upon the USDA's Office of International Cooperation and Development for advice and help. The OICD also assists the Agency for International Development as well as international agencies such as the Organization of American States and the Organization for Economic Cooperation and Development. It provides recommendations, shares technical knowledge, and cooperates with these organizations to help solve world food problems and to sponsor scientific exchanges and research. The OICD helps to improve American and foreign agriculture and promotes good relations with other countries.

Small Community and Rural Development

The Farmers Home Administration (FmHA), Rural Electrification Administration (REA), and Federal Crop Insurance Corporation (FCIC) are agencies that sponsor programs to support the development of small communities and rural areas. The FmHA provides credit, or loans, to rural residents, both farmers and nonfarmers, who are unable to obtain it from traditional lending institutions, such as banks and savings and loan associations. Farmers can use this money to purchase seed, equipment, and livestock, and to improve their land with

A farmer (left) reviews his loan application with a banker. Beginning farmers must borrow money to purchase seed, equipment, and livestock. The Farmers Home Administration offers loans to farmers when they are unable to obtain the funds from a bank.

A farmer checks the main control panel of his automatic poultry feed mill. The Rural Electrification Administration lends money to cooperative associations that need to construct power lines to provide adequate electric service to people in rural areas.

various soil-conservation techniques, or to repair homes, barns, or other buildings. Families with low incomes can also qualify for these loans to help them purchase suitable housing or to build or repair homes. Builders can receive FmHA loans to construct low-cost housing in rural areas for low-income residents. The FmHA also provides loans to individuals between 10 and 21 years of age to help them establish income-producing businesses, such as building and operating roadside fruit and vegetable stands.

The Rural Electrification Administration (REA) provides funds to assist rural electric cooperatives and telephone companies in furnishing those services to rural areas. REA funds are used by the cooperatives to construct electrical generating plants and to erect transmission and distribution lines for electric service to farms and rural residents. REA funds also help cooperative utilities and telephone companies provide service to isolated areas or to areas with low-population densities where the utilities' economic returns often do not meet their expenses.

The Federal Crop Insurance Corporation (FCIC) stabilizes agriculture by cushioning the effects of natural disasters that destroy farm crops. In 1987, farmers in 49 states could participate in the crop insurance program. Federal crop insurance helps to protect the farmer from losses due to bad weather (including hail, wind, rain, drought, hurricanes, tornadoes, floods, and lightning), insect infestations, plant diseases, fires, and earthquakes. Farmers who choose to participate in the crop insurance program must pay a nominal premium, or fee, in order to be covered, just as they pay premiums to private insurance companies for protection of their homes and automobiles from damages of various sorts. Farmers can insure their crops for 50, 65, or 75 percent of the yield. Suppose, for example, that a farmer had a corn crop insured for 75 percent of its yield and that his harvest had averaged 100 bushels per acre over the past several years. If he lost his crop to bad weather, he could receive payment for 75 bushels per acre. (Payments are based on current market prices.)

Staff Offices

The first of the four USDA offices that provide staff support is the Office of Budget and Program Analysis. This office coordinates the preparation of the department's budget and legislative reports by its component agencies. In addition, it analyzes USDA policy, programs, and budgets and helps the various agencies within the department review their regulations and devise new ones.

The second office advising the secretary and deputy secretary is the General Counsel. He or she offers advice on all legal matters and represents the USDA in courts of law and when testifying before Congress.

The third office supplying support to the secretary is the Office of the Inspector General. This office audits departmental funds to make sure that they are spent properly and to prevent wasteful expenditures. It is also responsible for detecting fraud (cheating) within the department. The Inspector General investigates the complaints of employees and provides physical security for the secretary at home and abroad. He or she is not appointed by the secretary but rather by the president, and the appointment must be confirmed by the Senate. This safeguard ensures the Inspector General's independence when investigating possible wrongdoing or when illegal actions might have been taken by an employee or agency.

The fourth office contributing to the administration of the department is that of the Judicial Officer. He or she makes the final decisions in the absence of the secretary regarding regulations and appeals of a judicial nature whenever a hearing is required by law to help solve a particular dispute. These cases are those that can be settled without going to court—for example, the resolution of a dispute between a farmer and the USDA over the boundaries of a national forest, the labeling of food, or a disagreement over whether a farmer had complied with certain conservation agreements in order to take part in an acreage diversion or price support program. Both parties must agree to accept the decision of the judicial officer before he or she makes a decision.

The Seven Division Offices

Seven assistant secretaries report to the secretary and deputy secretary in the following areas: Administration, Economics, Food and Consumer Services, Governmental and Public Affairs, Marketing and Inspection Services, Natural Resources and Environment, and Science and Education.

Administration

The Office of Administration consists of a board of contract appeals, an office of administrative law judges, an office of advocacy and enterprise, an office of finance and management, an office of information resources management, an office of operations, and an office of personnel. These offices are responsible for overseeing general personnel matters and ensuring the civil rights of and equal opportunity for all ethnic, racial, and religious groups represented by employees in the department. It supervises the safety and health programs that are presented to employees and handles the accounting of the USDA's financial concerns, including the purchase of equipment and land.

Economics

The Economic Research Service (ERS) provides economic information that will help officials formulate and administer the best programs for agriculture and rural affairs. In this respect, the ERS makes both short-term and long-term forecasts about American and world agricultural conditions. This information

A farmer wades through water in a field in the Red River Valley of North Dakota after a disastrous flood submerged hundreds of farms in the area. The Federal Crop Insurance Corporation provides insurance that covers unavoidable losses in crops caused by adverse weather conditions.

helps farmers and food manufacturers plan for the future and take advantage of the best possible prices, avoid surplus production, and meet food requirements at home and abroad. Indeed, this information is used by farmers, agricultural organizations, suppliers, marketers, processors, and consumers who make production, marketing, and purchasing decisions. Public officials in Congress, the state legislatures, and the counties also use this information to plan agricultural policy and programs. The ERS releases information in the form of printed materials, radio and television announcements, and through its computer, which provides direct access to farmers, suppliers, marketers, and legislators.

The National Agricultural Statistics Service (NASS) also prepares estimates and reports about agricultural production, supply and demand, and prices that aid the orderly marketing of agricultural commodities. These reports contain data on field crops, livestock, and processed food. The NASS surveys the prices paid for agricultural commodities, the prices received by farmers for their goods, and the amounts paid for labor. It prepares the reports by conducting a complex system of surveys involving farmers, ranchers, food manufacturers, and buyers through mail, telephone and personal interviews and field and factory visits. The agency also furnishes weekly, monthly, and annual reports for free distribution to the news media and to Congress.

The Economic Analysis staff helps the USDA develop, combine, and analyze statistical information for the formulation and evaluation of short- and intermediate-term agricultural policy—for example, precipitation deficiencies and corn, wheat, cotton, and soybean harvests—to determine whether policy adjustments should be made, such as allowing farmers to cut hay on lands specifically reserved for conservation during times of severe drought. This office also reviews the recommendations of other agencies in order to help shape agricultural policy, such as plans by the Soil Conservation Service and Forest Service to provide programs that will affect farmers. The Economic Analysis staff represents the USDA in meetings with industry and consumer groups that are held to discuss the economic effects of current and proposed USDA policies.

The Office of Energy handles all energy-related topics within the USDA. It works closely with the federal Department of Energy in determining energy activities that affect agriculture and rural America. Specifically, the Office of Energy estimates the amount of fuel—gasoline, diesel, and oil—that is required for agriculture on an annual basis. If allocation or rationing is needed, the office acts as a spokesman for the department.

The World Agricultural Outlook Board works to improve the reliability of all agricultural information originating from the department that is disseminated to the general public. This office helps gather information and interpret developments that will alter American and world agriculture and brings together interagency specialists to produce official estimates of supply, uses, and prices of products.

Food and Consumer Services

The Food and Nutrition Service (FNS) oversees programs that provide food to people in need. The department coordinates its work with state and local governments. The food stamp program, the major activity of the FNS, supplies food coupons to needy persons at low or no cost to increase their food purchasing power.

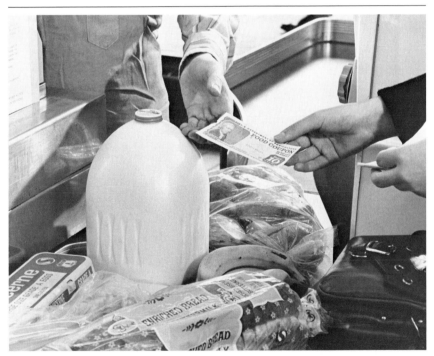

A customer uses food stamps to pay for groceries. The Food and Nutrition Service (FNS) issues stamps to thousands of low-income Americans, allowing needy participants to increase their buying power.

Elementary school children are provided with nutritious food for lunch by the FNS's National School Lunch program.

In 1964, Congress passed the Food Stamp Act to enable the USDA to use the nation's abundant foods "to the maximum extent practicable to safeguard the health and well-being of the nation's population and raise the level of nutrition among low-income households." To accomplish this goal, the USDA has issued coupons, commonly called stamps, to eligible households. Those who qualify for this program can purchase stamps that can be used to buy food at participating stores—any retail store that has been approved by the FNS to accept and redeem the coupons. Participants can purchase a $1 coupon for approximately $.85, thereby increasing their buying power. Grocery stores then turn the stamps over to the USDA for monetary reimbursement. Participants also receive "bonus" stamps to help provide an adequate diet. In 1970, Congress authorized the distribution of free food stamps to those people who could not afford to purchase them.

The FNS also administers the National School Lunch Program, the School Breakfast Program, and the Child Care Food Program to help provide meals to children in preschools and public and nonprofit private schools of high school level (with tuitions of $2,000 or less) that do not have the money to buy nutritious breakfasts and lunches. It also furnishes funds to the states to provide milk to children under the Special Milk Program, which is administered

86

in schools that do not participate in any other federal food program. The agency also provides food to women, children, and infants whose health is in danger because they cannot afford adequate food.

The Human Nutrition Information Service conducts research on the nutritional value of food and reports its findings to the public to improve the nutritional quality of diets. The Office of the Consumer Adviser represents the public to departmental policymakers and the department to Congress for consumer-related matters, such as concerns about the use of pesticides on fruits and vegetables or the health risks to the consumer of growth hormones or drugs used for the prevention of diseases in livestock. It also gives information to consumers about the services that the USDA supplies and evaluates consumer complaints.

Governmental and Public Affairs

The Office of Governmental and Public Affairs recommends policies that the USDA should review for possible adoption. It also coordinates the public information programs of the department. Staff have the responsibility to convey information between the department and Congress, to the mass communications media, such as newspapers and radio and television stations, and to state and local governments that need to know about USDA policies and programs.

Marketing and Inspection Services

The agencies responsible for marketing and inspection services are the Agricultural Cooperative Service, the Agricultural Marketing Service, the Animal and Plant Health Inspection Service, the Federal Grain Inspection Service, the Food Safety and Inspection Service, the Office of Transportation, and the Packers and Stockyards Administration.

The Agricultural Cooperative Service offers advice and organizational aid for cooperatives, such as conducting studies that will help cooperatives market farm products, purchase supplies, and operate efficiently. The Agricultural Cooperative Service also collects and publishes statistics and research about farm cooperatives in its monthly magazine, *Farmer Cooperatives.*

The Agricultural Marketing Service (AMS) is a departmental news service that compiles information about supply and demand, prices, location, and quality of agricultural commodities for radio, television, and newspapers each day and disseminates the information via a national satellite system. It also

The egg candling operation at a farm in Chesterfield, South Carolina. The eggs are passed over an intense light to test their internal condition for staleness, blood clots, and growth. The standards for grading eggs are established by the USDA's Agricultural Marketing Service.

develops standards, grades, and classification for more than 300 agricultural commodities, such as meat, poultry, eggs, fruits, vegetables, nuts, dairy products, cotton, and wool in order to certify the quality of these products. For example, the AMS establishes the grade standards for the size of eggs—Grade A small, large, and extra large. This grading system helps to assure the consumer that all of the eggs in a carton are uniform in size and it helps the grocer to determine a fair price. As a result, consumers pay more for large eggs than they do for small ones. The AMS also provides grading services to meat packers. Beef, for example, receives several grades, such as "choice." This helps the consumer select meat cuts based on the amount of marbeling, or fat, in the meat that affects its flavor and tenderness.

The AMS is also responsible for setting the guidelines for military and civilian agencies that purchase food and for overseeing fair marketing of fruits and vegetables, truthful advertising, and accurate labeling of seeds. The AMS inspects warehouses that store agricultural commodities and egg-processing plants to help protect the public's health. The AMS also establishes minimum prices that must be paid to farmers for the milk that they sell.

The Animal and Plant Health Inspection Service administers regulatory programs to protect and improve the health of animals and plants that are beneficial to humans and the environment. This agency is responsible for

The USDA's Grade A label for large eggs.

programs to eradicate pests and diseases from plants and animals and it imposes quarantines (enforced isolations to prevent the spread of disease or pests) and regulates the importation of foreign plants and plant products, animals and animal products, and birds that may harbor pests. The agency's staff oversees the health and humane treatment of all livestock and poultry when the animals are being shipped in interstate commerce. (Inspection services are located at all major ocean, air, border, and interior ports of entry in the continental United States and in Hawaii, Alaska, Puerto Rico, the Virgin Islands, the Bahamas, and Bermuda.) The agency also regulates the transportation, sale, and handling of circus and zoo animals as well as those animals used in laboratory experiments or for exhibition. Officials within the Animal and Plant Health Inspection Service assist farmers and ranchers in eradicating birds, rodents, and other predators to prevent the loss of crops and livestock.

The Federal Grain Inspection Service makes sure that the scales used to weigh grain are accurate, because the weight of grains such as wheat helps determine the price paid per bushel. The Service establishes standards for grading grain and administers a nationwide system of inspecting and weighing grain.

The Food Safety and Inspection Service checks meat and poultry products moving in interstate and foreign commerce for human consumption to ensure that they are wholesome and properly labeled. It inspects meat and poultry at the time of slaughter to assure that humane slaughtering techniques have been used. The service also checks the processing and handling of the products and inspects plant facilities and equipment for cleanliness.

The Office of Transportation devises transportation policy for agriculture and promotes the development and maintenance of efficient roads, highways, and railways. This office helps farmers and ranchers ship their commodities to market in a timely manner so that their goods will not spoil or be damaged.

The Packers and Stockyards Administration works to maintain fair competition and trading practices by food manufacturers. This agency also protects the consumer against meat processors who might try to monopolize or restrict the availability of meat and poultry in oder to increase prices.

Natural Resources and Environment

The Forest Service and the Soil Conservation Service are the two divisions of the USDA that handle matters in the areas of natural resources and the environment. The Forest Service, the largest agency in the USDA, protects the nation's national forests, helps to ensure the best methods for cutting

Gypsy moth larvae destroy a tree. The Forest Service works with state foresters to determine the best methods for exterminating pests and protecting wilderness areas.

timber, and assists state and private foresters in the areas of insect and disease control. The Forest Service manages 156 national forests, 19 national grasslands, and 17 land utilization projects that include 191 million acres of forest and grasslands. It has set aside 32 million acres as wilderness areas and 175,000 acres as primitive areas where timber cannot be harvested. The Forest Service is active in 44 states, the Virgin Islands, and Puerto Rico.

The Forest Service tries to sustain production of timber in accordance with the nation's demand for wood and paper products, to provide wildlife habitat, to protect water supplies, and to provide recreation areas, including picnic and camping grounds. Whenever trees are cut in national forests, for example, the Forest Service makes sure that the tree harvest will not damage the land, water resources, or wildlife habitat in the area and that reforestation will begin quickly.

In addition, the Forest Service conducts research to determine how timber, water, rangelands, wildlife, and recreation areas will best meet the needs of the public. At its National Forest Products Laboratory in Madison, Wisconsin,

White-tailed deer in their natural habitat in Michigan. The Forest Service protects wildlife, safeguards water supplies, and oversees logging in national forests to ensure that wildlife habitats are not damaged.

the Forest Service discovered a way to reduce stream pollution from wood-pulp manufacturing plants by recovering 90 percent of the chemicals used to make the pulp before they were discharged into nearby streams. The service's scientists have also worked on the development of new insect control methods that will eliminate the use of toxic sprays. In addition, the Forest Service has developed an electronic system that locates forest fires by using equipment in airplanes. An infrared scanner is used to pinpoint a "hot" spot and automatically mark the location on film. Meanwhile, a radar system determines the location of the airplane and records it on a computer. The information is compiled, enabling fire fighters to reach a forest fire in time to put it out before it spreads.

The Youth Conservation Corps and the Volunteers in the National Forests programs are two organizations that are directed by the Forest Service. These groups help to supervise campgrounds, to build roads and trails, to pick up litter, and to provide lifeguard services for recreation areas.

The Soil Conservation Service (SCS), the second agency that reports to the assistant secretary for natural resources and environment, has the responsibility for carrying out soil and water conservation programs to help farmers, other federal, state, and local agencies, and private corporations that use lands that are susceptible to erosion. The SCS works through locally organized and operated conservation districts, watershed protection projects (installations that retard soil erosion and prevent flooding), and resource conservation and development projects to improve the environment and to support community conservation programs. Approximately 3,000 conservation districts enable the SCS to offer expertise to farmers and other individuals in the United States, Guam, Puerto Rico, and the Virgin Islands. Since its creation in 1935, the SCS has helped protect more than 2 billion acres of land from erosion.

The SCS also assists state and local agencies in planning flood control programs in order to prevent water pollution and to safeguard high-quality

A science instructor explains the features of good wetland, or swamp-land, habitat to a group of students. The Soil Conservation Service supports community conservation programs and offers expertise to farmers on how best to combat soil erosion and water pollution.

A USDA soil scientist records information about the site he is survey-ing. The data he collects will be recorded on a soil map, which is used to assist the department in planning fertilization and conserva-tion programs: Scientists record the chemical composition of the soil, the location of water supplies, and the effects of wind erosion on the surface of the land.

water for rural areas. In addition, the SCS prepares soil surveys and maps to help farmers make the best use of their lands for conservation purposes.

Science and Education

Five agencies guide USDA programs in the areas of science and education—the Agricultural Research Service, the Cooperative State Research Service, the Extension Service, the National Agricultural Library, and the Office of Grants and Program Systems.

The Agricultural Research Service (ARS) engages in food and nutrition research and then tries to put the research to practical use. For example, the ARS scientists investigate the effects of the different kinds and amounts of fats and carbohydrates necessary for good nutrition. They also discovered that young men and women need larger amounts of magnesium in their diets than was previously believed (magnesium is a mineral that plays an essential role in the body's use of food).

94

ARS scientists are also involved in "utilization research," designed to develop and improve the uses of farm commodities in order to increase their value as food, feed, or industrial products. Utilization research can help create new markets for commodities, provide useful goods to consumers, and reduce surpluses of food and fiber, thereby enhancing prices. ARS staff have discovered methods to produce penicillin on a large scale, enabling the antibiotic to save millions of lives from pneumonia and other illnesses. Potato flakes—dehydrated mashed potatoes—were another important discovery made by the USDA; the department has licensed its method for making potato flakes to various food manufacturers. Scientists also discovered the process for making frozen orange-juice concentrate, a process used today by the major citrus juice manufacturers. Another example of USDA research was the discovery of a finish for wash-and-wear, or wrinkle-resistant, cotton fabrics that not only saves ironing time but also increases the use of cotton as a fabric for shirts and blouses. Today, USDA scientists conduct utilization research at five regional laboratories: Wyndmoor, Pennsylvania; Peoria, Illinois; Albany, California; Athens, Georgia; and New Orleans, Louisiana.

Scientists at the Mississippi State University experimental farm learn about the condition of a cotton crop from a computer screen. The computer simulates an entire growing season in minutes; the scientists check an automated weather station that feeds data from the cotton field into a farm computer, which in turn updates the cotton yield and harvest date. The Cooperative State Research Service gives financial assistance to state agricultural experiment stations to help them conduct research.

At the USDA's 193,000-acre Jornada Experimental Range north of Las Cruces, New Mexico, scientists in the Agricultural Research Service also conduct another form of research. There they are attempting to bond lambs to cows. In other words, they are trying to get lambs to think that cows are their mothers in order to protect them from coyotes that are out on the rangelands. Because cows protect their young while ewes (female sheep) do not, the scientists hope that the sheep population will not be so decimated by marauding coyotes. To accomplish this bonding, the scientists placed some lambs and a group of yearling heifers in a pen for 60 days. Then, they released the animals in a pasture. The cattle and sheep stayed close together, although they normally do not associate with each other on a common rangeland. After a period of time, coyotes had not killed any of the sheep, whereas another flock that had not bonded to cows suffered more than a 50 percent loss. If this research is successful, it will help ranchers prevent the loss of their sheep to coyotes and other predators and thereby increase their income from the sale of sheep for mutton or fleece for wool.

The second agency within the science and education division is the Cooperative State Research Service. This agency administers the funding for agricultural research that is conducted at the state agricultural experiment stations, colleges of veterinary medicine, the 1890 land-grant colleges, Tuskegee Institute (a college founded by Booker T. Washington in 1881 to train black teachers, and today a private coeducational school), and approved schools of forestry.

The Extension Service, the third agency in the division, is the education agency of the USDA and the federal partner of the Cooperative Extension System, a nationwide network (including the land-grant universities and more than 3,150 county offices representing local governments) that provides the results of agricultural research to the public. The Extension Service addresses major issues including methods to restore profitability in agriculture through improved farm and ranch management, methods to better manage soil, water, and natural resources, and methods to aid rural development and to improve family life. Its programs are designed to improve farmers' awareness of matters such as marketing, home economics, and nutrition, and include 4-H activities that help young people learn about career opportunities and develop leadership skills. (The 4-H program was created in 1900 to provide local educational clubs for the young—ages 9 to 19. It was designed to teach home economics and agricultural methods and to help improve the four H's: head, heart, hands, and health.)

The National Agricultural Library also falls within the domain of the science and education division. It is the largest agricultural library in the United States

Members of a 4-H club groom a calf. The USDA's Extension Service sends agents to agricultural organizations to teach members about improved methods of farm management, livestock care, home economics, and soil conservation.

and contains approximately 2 million books, including works on botany, chemistry, forestry, livestock, poultry, zoology, plant pathology, and general reference books on agriculture. Located in Beltsville, Maryland, the library serves USDA staff, federal agencies, land-grant universities, and the general public.

The fifth agency in the division of science and education is the Office of Grants and Program Systems. This office manages the program of competitive grants, or money that the scientists at the agricultural experiment stations, colleges and universities, and other institutions seek to support basic research projects. Basic research is often called pure research because it provides new knowledge that may not be immediately or directly applied to current agricultural problems. From this research, other scientists may conduct applied research to discover how best to use that knowledge, in practical terms, to aid farmers and rural residents.

97

A scenic view across Lake Malone in southwestern Kentucky, one of the lakes that the Soil Conservation Service helped to create as part of the Mud River Watershed—a basin designed to conserve water and soil.

SIX

The USDA's
Changing Heritage

In 1862, Congress charged the newly created Department of Agriculture with two far-reaching and important tasks: It was to conduct scientific research to discover new agricultural knowledge and to determine methods to apply that knowledge to everyday farming practices; and it was to disseminate this information to farmers across the nation so that they could increase their efficiency, productivity, and profits. Although agricultural research and education dominated the early history of the department, by the late 19th century the agency became increasingly involved in regulatory activities. The Department of Agriculture worked to prevent the importation of diseased livestock, to ensure the humane treatment of livestock shipped in interstate commerce, and to administer regulations concerning the purity of foods. The department's investigations of other food industry began with the oleomargarine manufacturers and later expanded to other food manufacturers, who sometimes used harmful chemicals for preservatives and often labeled their food products incorrectly.

During the early 20th century, the department increased its educational services with the passage of the Smith-Lever Act of 1914. This legislation created the Cooperative Extension Service, which sent agents and home economists into the counties to help farmers and homemakers apply the best practical agricultural and homemaking techniques to their daily lives. Although state and local governments contributed funds to this service, the extension service's personnel were joint employees of the federal government and the state agricultural universities.

The post–World War I agricultural depression caused farmers to demand other services from the Department of Agriculture. Many farmers wanted the department to take an active role in helping them locate new markets and expand old ones. Some farmers even wanted the federal government to purchase commodities to remove surpluses from the marketplace in order to keep prices at reasonable or profitable levels.

With the election of Franklin D. Roosevelt to the presidency in 1932, the Department of Agriculture became an activist agency, directly influencing the lives and finances of farmers on a daily basis. The Agricultural Adjustment Act of 1933 enabled the federal government to pay farmers to reduce production so that it could support profitable agricultural prices and reduce surpluses. Today, the USDA oversees a number of programs designed to keep prices at profitable levels and to support farm income—all for the purpose of helping the family farmer to avoid bankruptcy and to continue farming the land.

The Department of Agriculture continues to provide important services to the nation's farm, rural, and urban communities. The Forest Service and the Soil Conservation Service, for example, safeguard the precious natural resources of timberland, soil, and water for future generations. The agricultural experiment stations use USDA financial and technical assistance to increase scientific knowledge about farming, and the extension agents help share these discoveries with the American public. USDA scientists test new methods for using food products, investigate the safety of pesticides, develop new varieties of productive plants, prevent plant and animal diseases, and improve nutrition and the public's health.

The regulatory activities of the department assure fair agricultural prices for farmers and adequate food for consumers at reasonable costs. USDA programs also help to deter unfair trade practices in agriculture. By collecting and analyzing agricultural statistics at home and abroad, the department contributes to the shaping of public policy so that the federal government can respond to agricultural and rural problems and can improve the quality of rural and urban life. Other domestic responsibilities involve administering the food stamp and food assistance programs to schoolchildren and the needy so that they have adequate food and balanced diets.

The USDA represents the farmer in international trade agreements, helps developing nations by sending food and technical assistance to fight hunger and improve farm production, and trains foreign agricultural scientists. It also cooperates with other departments such as the Department of State and the

A scientist pours pesticide waste into the holding tank of the experimental mobile pesticide system at the USDA's research center in Beltsville, Maryland. The test's results will be used by researchers to help accelerate the breakdown of chemicals in pesticides that can pollute water supplies.

Department of Commerce in making international trade agreements that involve agricultural commodities.

Despite its past research, educational, and regulatory work, the USDA is not without its critics. Some people charge that the department's price support programs are too expensive and that the federal government should not pay farmers to curtail production. Others argue that the food stamp and food distribution programs encourage laziness on the part of those who participate in the programs. However, urban congressmen usually demand the retention of these programs in return for their support of agricultural legislation that provides price supports to farmers.

The USDA has also been criticized for supporting research that will help large-scale, wealthy farmers and the giant corporations that own land and are involved in agricultural production. These critics charge that the department, through its research facilities and the state agricultural experiment stations, does not conduct research that will enable the small-scale, family farmer to compete with the large-scale operations on an equitable basis. And, they have argued, the USDA-sponsored research in the area of technology often eliminates agricultural jobs.

Farmers also criticize the USDA for not doing enough to help the family farmer by means of the price support or loan programs handled by agencies such as the Farmers Home Administration or the privately controlled, but governmentally supported, Federal Land Banks. Others believe that the department could do more to expand markets for American commodities abroad.

Some critics of the department contend that it engages in work designed to make the farmer more productive while at the same time paying farmers not to raise or market certain crops. Some also argue that because tobacco products are dangerous to one's health, the USDA should not conduct research to make tobacco farmers more productive, or to encourage the use of tobacco products.

A USDA microbiologist prepares to inject chicks with an experimental vaccine designed to protect them from a disease carried by certain insect organisms.

Iowa farmers, faced with possible repossession of their machinery or land, gather together to protest the Farmers Home Administration's action to repossess a neighbor's machinery. The FMHA has been criticized for not meeting farmers' financial crises with increased price support and loan programs.

Certainly, the future will bring new concerns for the Department of Agriculture. The USDA will, however, continue to deal with production and marketing problems in an attempt to keep production in line with demand and to keep prices at adequate levels. But the department will probably give increasing attention to bioengineering—the process of creating or changing plant and animal genetic composition to develop new and more productive plants and animals that are also resistant to diseases. The public's concern about the safety of pesticides, herbicides, and fungicides will encourage USDA scientists to discover chemicals and natural controls that can be used safely to treat plants and animals but that will not harm humans or the environment. Besides trying to protect the use of the nation's land, the department will continue to fight rural poverty, to work to improve the quality of rural life, and to strive to end hunger in the world. Because it has such a history of service—more than 100 years—the Department of Agriculture has an important heritage upon which to build.

Department of Agriculture

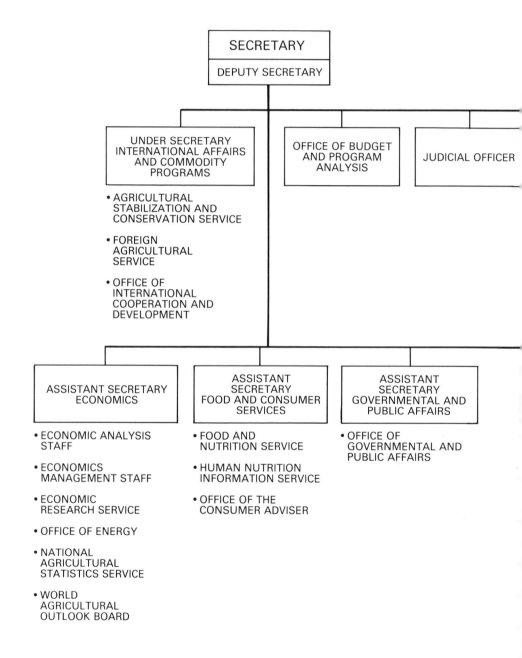

SECRETARY

DEPUTY SECRETARY

UNDER SECRETARY
INTERNATIONAL AFFAIRS
AND COMMODITY
PROGRAMS

OFFICE OF BUDGET
AND PROGRAM
ANALYSIS

JUDICIAL OFFICER

- AGRICULTURAL
  STABILIZATION AND
  CONSERVATION SERVICE

- FOREIGN
  AGRICULTURAL
  SERVICE

- OFFICE OF
  INTERNATIONAL
  COOPERATION AND
  DEVELOPMENT

ASSISTANT SECRETARY
ECONOMICS

ASSISTANT
SECRETARY
FOOD AND CONSUMER
SERVICES

ASSISTANT
SECRETARY
GOVERNMENTAL AND
PUBLIC AFFAIRS

- ECONOMIC ANALYSIS
  STAFF

- ECONOMICS
  MANAGEMENT STAFF

- ECONOMIC
  RESEARCH SERVICE

- OFFICE OF ENERGY

- NATIONAL
  AGRICULTURAL
  STATISTICS SERVICE

- WORLD
  AGRICULTURAL
  OUTLOOK BOARD

- FOOD AND
  NUTRITION SERVICE

- HUMAN NUTRITION
  INFORMATION SERVICE

- OFFICE OF THE
  CONSUMER ADVISER

- OFFICE OF
  GOVERNMENTAL AND
  PUBLIC AFFAIRS

OFFICE OF THE GENERAL COUNSEL	OFFICE OF INSPECTOR GENERAL	UNDER SECRETARY SMALL COMMUNITY AND RURAL DEVELOPMENT

- FARMERS HOME ADMINISTRATION

- FEDERAL CROP INSURANCE CORPORATION

- RURAL ELECTRIFICATION ADMINISTRATION

ASSISTANT SECRETARY MARKETING AND INSPECTION SERVICES	ASSISTANT SECRETARY NATURAL RESOURCES AND ENVIRONMENT	ASSISTANT SECRETARY ADMINISTRATION	ASSISTANT SECRETARY SCIENCE AND EDUCATION

- AGRICULTURAL COOPERATIVE SERVICE

- AGRICULTURAL MARKETING SERVICE

- ANIMAL AND PLANT HEALTH INSPECTION SERVICE

- FEDERAL GRAIN INSPECTION SERVICE

- FOOD SAFETY AND INSPECTION SERVICE

- OFFICE OF TRANSPORTATION

- PACKERS AND STOCKYARDS ADMINISTRATION

- FOREST SERVICE
- SOIL CONSERVATION SERVICE

- BOARD OF CONTRACT APPEALS

- OFFICE OF ADMINISTRATIVE LAW JUDGES

- OFFICE OF ADVOCACY AND ENTERPRISE

- OFFICE OF FINANCE AND MANAGEMENT

- OFFICE OF INFORMATION RESOURCES MANAGEMENT

- OFFICE OF OPERATIONS

- OFFICE OF PERSONNEL

- AGRICULTURAL RESEARCH SERVICE

- COOPERATIVE STATE RESEARCH SERVICE

- EXTENSION SERVICE

- NATIONAL AGRICULTURAL LIBRARY

GLOSSARY

Acreage allotment The amount of land that a farmer can use for crops and still be eligible for USDA price supports on certain crops.

Agricultural attaché A USDA employee stationed at an American embassy or consulate in order to assist American farmers and shippers in selling agricultural commodities abroad.

Commodity An article of commerce delivered for shipment; a product of agriculture or mining.

Disaster relief USDA payments to farmers to compensate them for crops lost as a result of natural disasters such as floods, hail, and wind.

Entomology A branch of zoology that deals with insects.

Food Stamp Program A USDA-run program that provides grocery coupons to low-income households to raise their "level of nutrition."

Land-grant College Agricultural and mechanical schools established by the Morrill Act of 1862, which provided for federal grants of land to the states for colleges to teach agriculture, engineering, and home economics.

Parity payment A USDA payment awarded to farmers when the price of their commodity falls below a certain level. It is intended to help farmers maintain a predetermined standard of living, comparable to that of Americans who work in areas other than agriculture.

Payment-In-Kind (PIK) A USDA program designed to limit production and reduce the stock of government-owned agricultural commodities by making USDA payments with government-owned crops. PIK is used mostly to pay farmers for refraining from growing a specific crop that is being over-produced.

Pesticide   A chemical agent used to destroy agricultural pests; because of various harvesting and processing techniques it may turn up in and contaminate some foods.

"Plant Explorers"   Employees of the USDA's Bureau of Plant Industry who search foreign countries for new species of plants that might be suited to American agricultural climates.

Price support   Artificial maintenance of prices at a predetermined level, usually through government action.

SELECTED REFERENCES

Baker, Gladys, et al. *Century of Service: The First 100 Years of the United States Department of Agriculture.* Washington, DC: United States Department of Agriculture, 1963.

Carrier, Lyman. "The United States Agricultural Society, 1852–1860." *Agricultural History* 11 (October 1937): 278–88.

Carstensen, Vernon. "Profile of the USDA—First Fifty Years." In *Lecture Series in Honor of the United States Department of Agriculture Centennial Year,* 3–17. Washington, DC: United States Department of Agriculture Graduate School, 1961.

Graus, John M., et al. *Public Administration and the United States Department of Agriculture.* Chicago: Published for the Committee on Public Administration of the Social Science Research Council by the Public Administration Service, 1940.

Harding, T. Swann. *Two Blades of Grass: A History of Scientific Development in the United States Department of Agriculture.* Norman: University of Oklahoma Press, 1947.

Hurt, R. Douglas. "The Poison Squad." *Timeline2* (February/March 1985): 64–70.

———. "REA: A New Deal for Farmers." *Timeline2* (December 1985/January 1986): 32–47.

Patrick, William. *The Food and Drug Administration.* New York: Chelsea House, 1988.

Rasmussen, Wayne D., and Gladys L. Baker. *The Department of Agriculture.* New York: Praeger, 1972.

Ross, Earle D. "The United States Department of Agriculture During the Commissionership: A Study in Politics, Administration, and Technology, 1862–1889." *Agricultural History* 20 (July 1946): 129–43.

Terrell, John Upton. *The United States Department of Agriculture: A Story of Food, Farms and Forests.* New York: Duell, Sloan and Pearce, 1966.

True, Alfred Charles. *A History of Agricultural Research in the United States, 1607–1925.* United States Department of Agriculture Miscellaneous Publication No. 251, 1937.

INDEX

Adams, John Quincy, 25
Agricultural Act of 1956, 63
Agricultural Adjustment Act (AAA), 53
Agricultural Adjustment Act of 1933, 16, 54, 100
Agricultural and Food Act of 1981, 69
Agricultural Enquiries on Plaster of Paris, 22
Agricultural Marketing Service (AMS), 87
Agricultural Research Service, 94
Agricultural Stabilization and Conservation Service (ASCS), 78
Agricultural Trade Development and Assistance Act, 62
Albany, California, 95
American Farm Bureau Federation, 53, 54
American Farmer, 25
American Society of Agriculture, 24
Arlington, Virginia, 44
Athens, Georgia, 95

Babcock, Stephen M., 39
Baltimore, Maryland, 25
Beltsville, Maryland, 44, 73
Bishop, William D., 28
Block, John R., 69, 70
British Food Mission, 60
Butz, Earl L., 68

Calhoun, John C., 27
Child Care Food Program, 86
Civil War, U.S., 31, 33
Clay, Henry, 25
Cleveland, Grover, 42

Colman, Norman J., 42
Columbia Agricultural Society, 24
Commerce, U.S. Department of, 101
Commodity Credit Corporation (CCC), 62, 64, 70, 71, 78
Congress, U.S., 22–24, 26–28, 35, 37, 39, 99
Coolidge, Calvin, 50
Cooperative Extension System, 16
Crawford, William, 25

Deere, John, 26

Ellsworth, Henry Leavitt, 25–27

Farmer Cooperatives, 87
Farmers Bulletins, 42
Farmers Home Administration (FmHA), 79, 80, 102
Farm Security Administration, 54
Federal Crop Insurance Corporation (FCIC), 79, 81
Federal Land Banks, 102
Food and Drug Administration, 50
Food and Nutrition Service (FNS), 85, 86
Food for Peace program (Public Law 480), 17, 62, 65, 73
Food Stamp Act (1964), 86
Forest Service, 19, 47, 90–92, 100
Foster, La Fayette, 31

Georgetown, Washington, D.C., 24

Soil Bank Program, 63
Soil Conservation and Domestic
    Allotment Act (1936), 54,
    100
Soil Conservation Service, 19,
    54, 56, 90, 93, 100
South Carolina Society for Pro-
    moting Agriculture and
    Other Rural Concerns, 22
Soviet Union, 67
Special Milk Program, 76
State, U.S. Department of, 100
Supreme Court, U.S., 54

Taft, William Howard, 44
Tama, Iowa, 44
Treasury, U.S. Department of
    the, 25, 37

United Arab Republic, 65
United States Agricultural So-
    ciety, 27, 28
United States Department of
    Agriculture (USDA)
    attachés, 16
    Bureau of Animal Industry,
        36, 37, 44
    Bureau of Chemistry, 47
    Bureau of Entomology, 47
    Bureau of Forestry, 47
    Division of Chemistry, 37
    division offices, 82–97
    establishment of, 15

expansion, 21–31
extension agents, 19, 49
Extension Service, 19, 49, 96,
    99
role in research and educa-
    tion, 16, 33–37, 39, 42–45,
    59
role in regulation, 15, 37, 53,
    54
role in conservation, 18, 19,
    47, 54, 56, 57, 90–93, 100
plant industry, 45
structure of, 75–97

Volunteers in the National For-
    est, 93

Washington, George, 22–24
Washington, D.C., 24, 34
Wallace, Henry A., 53, 54
Wallace, Henry C., 50, 53
Watson, Elkanah, 25
Wesleyan University, 37
Whitney, Eli, 22
Wiley, Harvey W., 37
Wilson, James, 42, 44, 47
World War I, 49, 60
World War II, 57, 59
Wyndmoor, Pennsylvania, 95

Young, Arthur, 22
Youth Conservation Corps, 93

R. Douglas Hurt is associate director of the State Historical Society of Missouri in Columbia, Missouri. He holds a Ph.D. in American history from Kansas State University and is the author of *The Dust Bowl: An Agricultural and Social History, American Farm Tools: From Hand-Power to Steam-Power,* and *Indian Agriculture in America: From Prehistory to the Present.*

Arthur M. Schlesinger, jr., served in the White House as special assistant to Presidents Kennedy and Johnson. He is the author of numerous acclaimed works in American history and has twice been awarded the Pulitzer Prize. He taught history at Harvard College for many years and is currently Albert Schweitzer Professor of Humanities at the City College of New York.